PRINCIPLES OF QUANTUM MECHANICS

PRINCIPLES OF QUANTUM MECHANICS

by

ALFRED LANDÉ
Professor of Physics in the Ohio State University

CAMBRIDGE
AT THE UNIVERSITY PRESS
1937

CAMBRIDGE UNIVERSITY PRESS
Cambridge, New York, Melbourne, Madrid, Cape Town,
Singapore, São Paulo, Delhi, Mexico City

Cambridge University Press
The Edinburgh Building, Cambridge CB2 8RU, UK

Published in the United States of America by Cambridge University Press, New York

www.cambridge.org
Information on this title: www.cambridge.org/9781107667839

First published 1937
First paperback edition 2013

A catalogue record for this publication is available from the British Library

ISBN 978-1-107-66783-9 Paperback

CONTENTS

PREFACE

It is the aim of this book to develop the principles of quantum mechanics on the basis of a few standard observations. In this way we hope to succeed in eliminating from the customary interpretation of the theory some unphysical ideas that have no counterpart in empirical facts. Such a task would be quite trivial in the case of classical mechanics whose path, from the eighteenth century down to Einstein's relativization of the absolute time, was marked by a gradual elimination of anthropomorphic concepts. In quantum mechanics, however, only twelve years have passed since this theory was introduced as a cryptic technique of mathematical operations with non-commutative quantities, corresponding to a still more mysterious behaviour of matter whose particles seemed to disregard the laws of mechanics in favour of wave rules. But in spite of the Heisenberg principle of uncertainty which clarified so much of the physical content of quantum mechanics, and in spite of the mathematical perfection of the theory, there seems still some work to be done before the interpretation of the formulae satisfy all requirements of consistency. In this respect we can learn a great deal from the general theory of relativity, as may be seen from the following list of analogies between relativity and quantum mechanics.

(*a*) In Einstein's theory one describes a phenomenon, for instance, the motion of a falling stone, in two equivalent ways: one either derives the curved path of the stone from the forces of gravity, or ascribes it to the inertia with respect to an accelerated frame of co-ordinates. There is then a unique mathematical relation between the two descriptions.

(*b*) It would contradict however the very idea of relativity if one should apply the concepts of both descriptions *simultaneously* —that is, if one should ask for the distribution and magnitude of gravitational forces *within* the accelerated frame.

(*c*) On the other hand, the two explanations of the curved path are *equivalent*—that is, there is a unique mathematical connection between them. The same coefficients $g_{\mu\nu}$ that play the role of coefficients of force in the one interpretation, appear to be metric coefficients in the second interpretation. Since, however, these coefficients g, by virtue of their metric character, have to obey certain inherent differential equations, the same differential equations hold then for the g's in their first role as coefficients of force. In this way Einstein found his fundamental equations of the gravitational field.

Similar considerations apply now to quantum mechanics also:

(*a*) In quantum theory we can explain one and the same phenomenon, for instance the diffraction of light at a grating, in two *equivalent* ways: either we ascribe the diffraction pattern to a periodic distribution of matter in the diffracting instrument, which serves as a Huygens source of secondary interfering light *waves*; or we explain the same diffraction with the help of directed impulses imparted to the incident *photons* by corpuscles of matter supposed to be present in the diffracting apparatus.

(*b*) It would contradict however the basic idea of quantum mechanics if we should apply both ideas *simultaneously*, e.g. if we should inquire for the location of the impulse giving particles of matter and attempt to place them mainly in the beat maxima of the periodic waves of matter (since the latter were introduced hypothetically only for the purpose of explaining the diffraction with the help of light waves). Forgetting that equivalence precludes simultaneity has often led to paradoxical situations in the theory of quanta, for instance, to the question of how a photon is able to know which way it has to go, after having passed through a periodic grating. The answer given first by P. Duane is that from the standpoint of photons the diffracting instrument is not a periodic grating but is an arrangement of matter that gives off momentum only by amounts which are multiples of a certain basic amount.

(*c*) On the other hand, the two explanations of the diffraction pattern with the help of waves and corpuscles are *equivalent*—that is, there is a mathematical relation between the wave data and the corresponding corpuscular data. The most familiar of these relations are the Planck formula $\epsilon = h\nu$ and the de Broglie formula $p = h/\lambda$. It is the aim of the methods of Schrödinger and of Heisenberg-Born-Jordan to connect the wave description with the corpuscular description of the same phenomena in a general way. Now since the waves have certain inherent features—for instance, since a standing wave can have only an integral number of nodes, the same feature will have a bearing on the corresponding feature of corpuscles; their energies and momenta appear to be "quantized". It would, however, be misleading to ask, *within* the frame of the wave picture, for the location of particles. The concepts "waves" and "corpuscles" are complementary, but they cannot be applied simultaneously. If one does it, one commits the same error as a certain book of military instruction which tells us that a bullet falls down for two reasons, firstly because of the attractive force of the earth, secondly because of its own heaviness (meaning probably its inertia in an upwards accelerated system).

In view of the complementarity of the two classical theories and in order to avoid a confusion of their respective ideas, it is a good policy to compare the tentative theoretical explanation of an observed process with a complementary interpretation in which the roles of particles and waves have been interchanged. If the latter interpretation turns out to be objectionable, one is prepared to criticize the former interpretation also, even though it may represent a generally accepted opinion. A similar programme of exploiting the perfect complementarity of the ideas of corpuscles and waves in interpreting the observed facts has been carried through in Heisenberg's University of Chicago lectures on "The Physical Principles of Quantum Theory" (1930), a standard work to which the author is very indebted. The aim of the present book is different from Heisenberg's in that more stress is laid on the mutual dependence of the various principles of quantum theory, and on developing them from the simple theory of observation which leads eventually to the transformation theorems of P. Jordan.

Considerations of relativistic invariance together with Dirac's theory of spinning electron have been omitted from this book. Nor was it convenient to discuss the applications of Pauli's exclusion principle, which is quite alien to the interwoven system of the other principles of quantum mechanics.

It is a pleasure to express my gratitude to my colleagues Jerome B. Green and George H. Shortley for their great help in revising the manuscript, and to the Cambridge University Press for the beautiful typographical work as exemplified in the following pages.

A. L.

April 1937

INTRODUCTION

§1. OBSERVATION AND INTERPRETATION

The discussion of the physical nature of matter must be based, in the last analysis, on measurements. Such measurements aim at observable qualities such as the distribution of matter in space, its energy and momentum, its electric charge—in short its intensity. These measurements cannot be made, however, until we have found a tool, a measuring instrument whose reaction to matter is to be observed, and which itself is decidedly different from, and simpler than, the matter to be observed; so we use light, which is emitted, absorbed, reflected, refracted by matter and gives us information about matter itself.

Any such information implies, however, an hypothesis concerning the constitution of the light itself and its interaction with matter. There have been in fact *two* such hypotheses: first, the undulatory theory, that light consists of waves originating in, or modified by, matter; second, the corpuscular theory, that light consists of photons reacting with matter and behaving according to the mechanical rules of the conservation of energy and momentum. Two such antagonistic hypotheses, used for the interpretation of the same optical phenomena, result in two quite different views of the observed object, in *two different sets of data* describing the observed piece of matter. The wave hypothesis of light leads us to a distribution (also hypothetical) of matter in space, while the corpuscular theory of light will give us information about various amounts of energy and momentum apparently carried by the piece of matter. It is not surprising that these two sets of data, obtained by the two antagonistic theories of light, are not capable of being fused into one consistent mechanical model of matter. Nevertheless, quantum theory shows that the two sets of data are "complementary", not only from the standpoint of the two interpretations of optical observations, but also as inherent in matter itself. There are direct *formal mathematical relations* between the one set of data and its complementary set. For instance, there is a mathematical relation $E = h\nu$ between the changes of the *energy* E carried by matter (as judged from the corpuscular theory of light) and the *frequencies* ν of apparent density vibrations (as judged from the wave theory of light).

The main object of quantum mechanics is to develop the direct mathematical relations between complementary data after discussing their origin and their consequences from a physical point of view.

§ 2. DIFFICULTIES OF THE CLASSICAL THEORIES

(*a*) Let us review first some of the difficulties confronting the classical theories of matter. This will prepare us for abandoning the classical theories later on or at least for considering their concepts with greater scepticism. Since the beginning of this century we have known that matter is bound inseparably to electric charge. In particular, the anode and cathode rays with their deflections in electric and magnetic fields, and the tracks of radioactive rays in a Wilson cloud chamber, give us first-hand information about a corpuscular structure of matter. That is, down to dimensions as small as 10^{-12} cm. every volume element of space is either empty or contains the charge

$$e = 4{\cdot}77 \times 10^{-10} \text{ e.s.u.}$$

and the mass $m = 0{\cdot}9 \times 10^{-27}$ gram (electron) or $M = 1{\cdot}66 \times 10^{-24}$ (proton) or integral multiples of them.

(*b*) Upon this corpuscular theory a hydrogen atom would seem to consist of a positive proton and a negative electron revolving around their common centre of gravity (Rutherford). The difficulty with this dynamical model of charged particles is that it conflicts with the electromagnetic theory of radiation, since the mechanical energy of revolution should be gradually converted into energy of radiation, and the electron should pursue a spiral curve into the proton with ever-increasing frequency of revolution. The model should emit then a continuous spectrum of frequencies instead of a series of separate spectral lines as observed.

There remains a possibility of saving the dynamical model of Rutherford by abandoning the classical theory of electromagnetic radiation in favour of a mechanical theory of light as presented in Einstein's theory of photons. Thus, according to Niels Bohr an atom was assumed to exist only in certain stationary states with certain selected energy values, and the emission of a photon of the energy ϵ was connected with a decrease (jump) of the atomic energy from a value E_m to E_n, releasing the balance $\epsilon = E_m - E_n$. This was supposed to happen in contradiction to the rules of the electromagnetic wave theory of light. And yet this sacrifice, made in order to save the mechanical model of the atom, was supplemented by an encroachment into the very mechanical model itself. The orbits were confined to "quantized" orbits selected by certain quantum conditions. One may say then that the original Bohr theory modified the mechanical model of the atom only slightly by quantum conditions, but abandoned the electromagnetic theory of light altogether. And yet, in the case of more than one electron (He, Li, etc.) the Bohr theory gave only approximately correct results.

(*c*) One may start just as well from the opposite point of view, that of retaining as much as possible of the electromagnetic wave theory of light. An atom, considered as the source of emitted, absorbed, or diffracted light waves, would then suggest a quite different picture: the atom would appear to contain a cloud of negative charge, the bulk of it being condensed within 10^{-8} cm. from the nucleus, but shading off with ever-decreasing density to infinity, the total charge being e, and the relative density being a continuous function $\rho\,(xyzt)$, where $\int \rho\, dv = 1$.

A closer study by L. de Broglie (1) and E. Schrödinger (2) revealed that the *density* function $\rho\,(xyzt)$, which describes the distribution of the total charge e in space and time, conforms with the intensity of certain vibrations or waves. In particular, there are *standing waves* in space that have either *one* or *two* or, in general, n nodes and n loops. Their amplitudes $\psi_n\,(xyz) \,.\, e^{2i\pi\nu_n t}$ give rise to charge densities $e \,.\, \rho_n\,(xyz) = e\,|\,\psi_n\,(xyz)\,|^2$ in the 1st, 2nd, ... nth stationary states of the atom, if judged from the electromagnetic wave theory.

(1), (2), etc., Index of literature on p. 117.

Since the loops and nodes of a standing vibration $\rho_n(xyz)$ do not change their places in time, they do not give rise to emission or absorption of electromagnetic light waves, but only to stationary polarization effects. In this way one understands that the "stationary states" do not gain or lose energy by radiation.

On the other hand, there seem to be states where *two* standing waves interfere with one another. If, for instance, the matter wave $\psi_n(xyz).e^{2i\pi\nu_n t}$ is superposed on the wave $\psi_m(xyz).e^{2i\pi\nu_m t}$, they produce a *beat* whose density $\rho_{mn}(xyzt)$ has $m-n$ beat maxima and minima in space which change their place in time with the frequency $\nu_m-\nu_n$. According to Maxwell's theory the density ρ_{mn} gives rise to the emission of electromagnetic waves whose frequency ν_{mn} is likewise $\nu_m-\nu_n$. In this way, Schrödinger was able to explain the frequencies ν_{mn} of observed spectral lines in accordance with the combination principle $\nu_{mn}=\nu_m-\nu_n$ of Rydberg and Ritz. This result is achieved, however, only by dropping the mutual reaction of the various volume elements of the charge cloud. Thus the theory deviates from what one would call an ordinary wave theory of charged matter. At the same time the theory abandons the view that the charge e is condensed in corpuscular electrons subject to the rules of mechanics.

(*d*) In order to save the corpuscular theory of electrons, in spite of its inconsistency with the electromagnetic wave theory, the following compromise has been suggested by M. Born (3). Consider the de Broglie-Schrödinger density ρ only as the "time exposure" of a corpuscular electron during its motion through space. Let $\rho(xyz).dv$ mean the *probability* of finding the electron e in volume elements dv at various places. The total probability is $\int\rho dv=1$. Or put in statistical terms: if a large number N of hydrogen atoms in the same stationary or transitory state are present, then they represent N electric dipoles with such a statistical distribution of their dipole moments and directions, that if they were crowded together they would result in the same charge cloud as N protons with their Schrödinger clouds crowded together.

The N Born dipoles represent the same total electric moment as do the N Schrödinger charge clouds. Thus both will give the same optical effects according to the wave theory of light. Furthermore, one understands now why the various volume elements of the density cloud do not repel one another: they are charged successively, not simultaneously.

(*e*) Now since in a state of transition $m \rightarrow n$ the Schrödinger clouds $\rho_{mn}(xyzt)$ change periodically in time, the corresponding dipoles of Born must be kinematic models. They will be static only when ρ_{mn} is constant in time, that is in the case of a stationary state $m = n$. If one tries however to carry out the redistribution of the N Schrödinger charge clouds into N Born dipoles at successive times, one cannot make the kinematic or static dipoles comply with the laws of dynamics, i.e. with the equations of motion of the electron in the Coulomb field of the proton. For instance, an electron in a state of given negative energy E_n should never be found outside the maximum distance of $r = \dfrac{e^2}{|E_n|}$ from the nucleus, for otherwise its kinetic energy would have to be negative. But the Schrödinger cloud has a finite density at all distances from the centre. Secondly, in a stationary state where Schrödinger's ρ is constant in time, each of the corresponding dipoles ought to be static, in contradiction to the force existing between its two poles. Thus, the statistical interpretation of the density cloud, introduced in order to save the theory of mechanical corpuscles, leads straight into new contradictions to mechanics.

(*f*) The same must be said of the attempt by Schrödinger himself to reconcile the corpuscular theory of electrons with his matter waves. A short signal can be said to represent the superposition of a number of monochromatic wave components, each of them extending over the whole of space, but annihilated by mutual interference except for a small range of space. Vice versa, one can build up a high maximum of wave intensity in an extremely small volume by superposing a great number of different waves with suitable amplitudes and phases. Schrödinger's idea was that what we take for a corpuscle is only the high crest or beat maximum of such a group of waves. The path along which a group maximum would travel according to the rules of the wave theory can indeed be proved to coincide with the path of a particle which travels according to the rules of dynamics. The snag in this wave interpretation of corpuscles is found in the fact that the beat maximum of waves in a dispersing medium will flatten out gradually; and the steeper it was originally, the faster it will flatten. So these corpuscular maxima would blur out within a very short time if the rules of the wave theory are applied to them—just as the Born dipoles would collapse if they were subject to the rules of dynamics. Yet we cannot admit that Schrödinger's wave interpretation of corpuscles is inferior to the now generally accepted corpuscular statistical interpretation of the wave intensity. Particles guided by the rules of waves are just as obscure as wave beats kept together by corpuscular postulates.

§ 3. THE PURPOSE OF QUANTUM THEORY

Quantum theory starts from a critical review of the foregoing contradictions, in particular, however, from the positive remark that optical signals coming from matter can always be interpreted both in terms of the wave theory and in terms of the corpuscular theory of light. If wave optics is used, one will ascribe wave properties also to matter, such as frequency, phase, amplitude, and intensity. If the corpuscular theory of light is applied, one will see corpuscular properties also in matter, such as kinetic and potential energy, momentum, the probability of their changes from one to another value. Although there is no possibility of fusing these two pictures into one unique image, it is all the more significant, however, that there is a direct relation between corresponding wave and corpuscular quantities, each being "complementary" to the other. The most familiar example of this relation is the formula $E_m - E_n = h\nu_{mn}$ relating the corpuscular energies of an atom before and after the emission of a corpuscular *photon* to the frequency ν_{mn} of the vibration in the charge cloud ρ_{mn} of the same atom if considered as the source of light *waves*. Quantum theory gives the general mathematical method of finding the relation between given corpuscular quantities and the complementary wave quantities and vice versa. We can develop this mathematical method of quantum theory by discussing a number of standard examples (Part I) and generalizing the results gained in these simple cases. We shall learn for instance how one calculates the periodic changes along a matter ray (its "wave length") as judged from the wave theory of light, if the same matter ray, as judged from its reaction to photons, appears to contain corpuscular momenta $+p$ and $-p$ of opposite directions. The most interesting results of quantum theory are obtained, however, in those cases where by virtue of inherent peculiarities of the problem there are only certain selected states possible in terms of the wave theory—and consequently in terms of the corpuscular theory too. For instance, the wave theory allows a set of standing waves that have either 0 or 1 or 2 or in general n nodes, where n is necessarily an integral number. If the same phenomenon is described afterwards in corpuscular terms, there will be only a selected series of mechanical energy values E_0, E_1; ... E_n, ... complementary to those standing waves. Intermediate energy values, although allowed mechanically, do not occur on account of that complementarity of corpuscular to wave quantities. It would be wrong, however (although it has become customary), to say that the continuous density function ρ_n stands only for the probability of finding real corpuscular electrons at various places. On the contrary, the continuous charge density ρ_n with its frequency ν_n is just as real from the wave point of view as are states with energy values E_n from the corpuscular point of view. Irrational are those customary attempts to fuse both aspects in one *image*, like "corpuscles that follow the rules of waves" or "waves whose vibrational energy is confined to, and changes by, quantised amounts".

Quantum theory claims to give a general method for translating corpuscular data into wave data and vice versa. Its purpose is not to explain or to fuse contradictory concepts. To give another example: A beam of matter or light always displays a certain amount of fluctuation in intensity, depending on the absolute intensity and on the homogeneity of the beam, that is, on its definiteness of colour and direction. If we ask why the fluctuations have that magnitude, meaning a reduction to one of the familiar pictures, the tentative answer might be that light consists of particles or perhaps of waves. The observed fluctuations, however, do not agree with either of these explanations. A shower of mechanical particles would explain the observed fluctuations only in the case of a small intensity of the beam. Interference of waves would suffice only in the limit of large intensity. But if instead we ask only "how" or what are the observed magnitudes of the fluctuations, and whether there is a mathematical method for calculating them for every given intensity, then quantum theory gives the answer.

There are two mathematical forms of the theory of quanta—the *wave mechanics* of Schrödinger (2), which starts from the wave concept, and the *quantum mechanics* of Born, Heisenberg (4) and Jordan (5) (and of Dirac (6)), which is closer to the corpuscular aspect. Both are equivalent, however, in their physical results. We take pains in this book to emphasise the perfect complementarity of the wave and the corpuscular picture and to contrast every statement made in wave language to a complementary statement expressed in corpuscular terms.

PART I

ELEMENTARY THEORY OF OBSERVATION (PRINCIPLE OF COMPLEMENTARITY)

§4. REFRACTION IN INHOMOGENEOUS MEDIA (FORCE FIELDS)

It is of particular importance for the understanding of quantum mechanics to realize that there are a number of phenomena that can be explained just as well by means of the *wave theory* as by means of the *corpuscular theory* of matter. These phenomena and their twofold interpretation will help us derive the simplest formulae of quantum theory and so become familiar with that peculiar *complementarity* of N. Bohr (7) of waves and corpuscles which will enable us later to deal with more complicated phenomena.

Suppose a beam of matter to travel along a certain curved path. In order to explain its deviation from the straight line the *corpuscular theory* would assume that a force field is acting on the particles of the beam changing their kinetic energy $K=\frac{1}{2}mv^2$ at the expense of their potential energy $U(xyz)$, so that the total energy $E=K+U$ remains constant. For a given total energy E, a particle at the point xyz will have a momentum $p=mv$ given by

$$p(xyz)=mv=\sqrt{2m\,.\,\tfrac{1}{2}mv^2}=\sqrt{2mK}=\sqrt{2m\{E-U(xyz)\}}.$$

The path of a particle of given energy E between two points A and B in the field of the potential energy $U(xyz)$ is always such that the line integral

$$\text{(1)}\qquad \int_A^B p(xyz)\,ds=\int_A^B \sqrt{2m\{E-U(xyz)\}}\,ds=\text{minimum}$$

along this path is smaller than the integral over any other line joining A and B. This is the "principle of least action" of Maupertuis; the minimum condition (1) is sufficient to determine completely the curved mechanical path between A and B in the field U.

The same curved path can, however, be explained just as well by means of the *wave theory*. One may assume that the beam consists of waves of constant frequency ν travelling through a medium with a variable index of refraction $n_\nu(xyz)$, so that the wave length $\lambda = \dfrac{\lambda_\infty}{n_\nu(xyz)}$ varies from point to point, λ_∞ being the wave length in a region where $n_\nu = 1$. The principle of Fermat says that the actual beam between two points A and B in the medium n will be that one for which the integral

$$(2) \qquad \int_A^B n_\nu(xyz)\,ds = \int_A^B \frac{\text{const.}}{\lambda(xyz)}\,ds = \text{extremum},$$

as compared with the integral over any other line joining A and B.

Comparing (1) with (2) one sees that the two theories (particles of constant E or waves of constant ν) account for the same curved beam if

$$p(xyz) = \frac{\text{const.}}{\lambda(xyz)}, \quad \text{or} \quad n_\nu(xyz) = \frac{\sqrt{E - U(xyz)}}{\sqrt{E}}.$$

That is, in order to explain the same curved beams the wave theory (geometrical optics) requires at every point (xyz) a wave length λ which is inversely proportional to the mechanical momentum p needed in the corpuscular theory for the explanation of the same beam. Although both λ and p change from point to point, their product must be supposed to be constant:

$$(3) \qquad p(xyz) \cdot \lambda(xyz) = \text{constant}.$$

This fundamental formula establishes a relation between the corpuscular momentum p and the wave length λ attributed to the same beam.

From the curved path of a beam neither the absolute value of p nor the absolute value of λ can be determined. Indeed, the same curved beam would result from k times as large a momentum p, if at the same time $\sqrt{E}$ and $\sqrt{U(xyz)}$ be assumed to be k times as large. And the same curved beam would result also if λ were assumed to be k times as small, if at the same time the refractive index $n_\nu(xyz)$ were assumed to be k times as large at every point of the space.

Both the absolute values of p and λ can be determined only by additional experiments in which these values are measured relative to certain gauges (§ 7). The magnitude of the constant in (3) will then turn out to have the universal value of Planck's h.

Fermat's principle (2) holds only as long as the radius of curvature of the beam is everywhere above a certain limit, that is, as long as the index of refraction has not too large a gradient. Else we are faced with deviations from geometrical optics, as the beam undergoes a diffraction into various directions at the same time. It is this lower limit of $1:\mathrm{grad}\, n$ which defines somewhat vaguely a new characteristic of the beam, its absolute "wave length". The familiar diffraction experiments are only a more definite way of measuring that inherent wave length λ of a beam in comparison with the known dimensions of an artificial or natural periodic grating constituting a periodic set of inhomogeneities in space.

On the other hand, the same ray displays a certain absolute value of its corpuscular momentum only when deviations from its path of constant energy E are observed, for instance in processes of collision. The absolute value of the mass m can then be determined relative to the known mass M of (large) objects with which energy and momentum are exchanged.

§ 5. SCATTERING OF CHARGED RAYS

If protons are travelling past a heavy nucleus $+Ze$, they will be deflected in various directions, the angle of deflection depending on their energy E and on the distance of their initial rectilinear path from the nucleus. The statistical angular distribution of the scattered protons complies with Rutherford's classical formula as long as the initial velocity v_∞ is small compared with c, in particular for

$$\frac{v_\infty}{c} \ll Z\,\frac{2\pi e^2}{hc} = Z\,\frac{1}{137}.$$

One would obtain exactly the same curved paths of proton-*waves* if one assumed that the nucleus produces an index of refraction

$$n(r) = \frac{\lambda_\infty}{\lambda(r)} = \frac{p(r)}{p_\infty} = \frac{\sqrt{E-U}}{\sqrt{E}},$$

where λ_∞ is the original length of the proton-waves. Fermat's principle applies, however, only if $\lambda(r)$ is everywhere small compared with the radius of curvature, that is if λ_∞ is small compared with the semicircle $\pi r_{\text{min.}} = \pi . \frac{Ze^2}{E}$. This condition is identical with the former condition for v_∞, since $E = \frac{m}{2} v_\infty^2$ and $p . \lambda = h$. Since $n(r)$ depends only on the ratio $\frac{\lambda_\infty}{\lambda} = \frac{p}{p_\infty}$, the scattering of fast protons can tell us nothing about the *absolute* value of their momenta or their wave lengths.

One can derive a scattering formula by means of a strict wave equation

$$\Delta\psi + \frac{4\pi^2}{\lambda^2}\psi = 0 \quad \text{with} \quad \lambda = \frac{\lambda_\infty}{n(r)}$$

for the wave amplitude $\psi(xyz)$. Assuming that ψ is the superposition of incident waves $\psi^{(0)} = e^{2i\pi x/\lambda_\infty}$ and perturbed scattered waves $\psi^{(1)}(xyz)$, one may solve the wave equation by means of a perturbation method as long as $\psi^{(1)} \ll \psi^{(0)}$. The latter condition is satisfied under the same conditions for λ_∞ as before. In this approximation one finds again Rutherford's scattering formula.

§ 6. REFRACTION AND REFLECTION AT A PLANE

Refraction (and reflection) can be explained by means of particles which preserve their energy, or by waves which preserve their frequency, when passing from one medium to another. The index of refraction n for waves changes discontinuously, producing an infinitely small radius of curvature which furnishes no gauge for measuring the absolute value of the wave length λ. Nor can one determine the absolute value of the momentum p of particles before and after their deflection in the infinite force field that has to be supposed in the transition layer. But the general rule (3) applies again.

Indeed, the *wave theory* assumes that the incident and the refracted (and reflected) waves

$$\sin\left(2\pi\nu_k t + \frac{x\cos\alpha_k + y\sin\alpha_k}{\lambda_k} + \delta_k\right) \qquad \begin{array}{l} k=1 \text{ incident,} \\ k=2 \text{ refracted,} \\ k=3 \text{ reflected,} \end{array}$$

have the same constant phase difference at all times all along the refractive plane, that is, for every value of t and x on the plane $y=0$. Hence we have

$$(4) \quad \nu_1=\nu_2 \quad \text{and} \quad \frac{\cos\alpha_1}{\lambda_1}=\frac{\cos\alpha_2}{\lambda_2},$$

that is, the rule of refraction—and

$$\cos\alpha_1=\cos\alpha_3 \quad \text{or} \quad \alpha_1=\alpha_3,$$

that is, the rule of reflection.

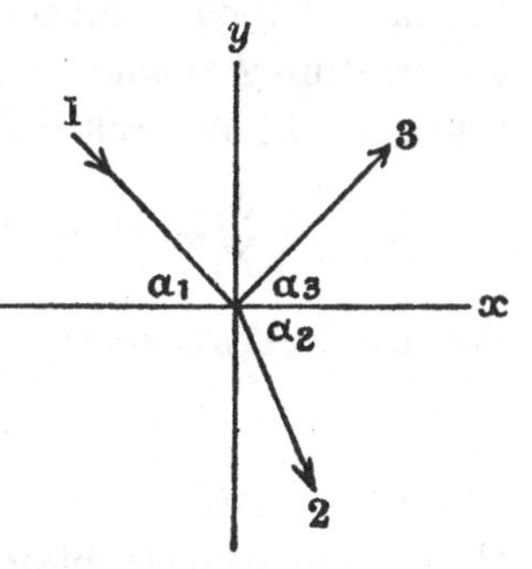

Fig. 1.

If, instead, one assumes incident *corpuscles* of the momentum p_1 being subject within the plane $y=0$ to forces parallel to the $\pm y$ axis, then the x-component of their momentum keeps constant:

$$(4') \qquad p_1\cos\alpha_1=p_2\cos\alpha_2 \quad \text{(conservation of } p_x\text{)}.$$

Both (4) and (4′) express the same rule of refraction if the relation (3),

$$p_1\lambda_1=p_2\lambda_2=\text{constant},$$

is valid.

Rules concerning the ratio between refracted and reflected *intensity* are derived from boundary conditions for waves. They depend on the special kind and polarization of the waves. Intensity rules for corpuscular rays when passing through the plane $y=0$ may be obtained by supposing that the particles obtain impulses in the two opposite normal directions $\pm y$ in such a numerical ratio that the plane itself neither gains nor loses momentum on the average.

§ 7. ABSOLUTE VALUES OF MOMENTUM AND WAVE LENGTH

Suppose a beam of matter to be bent by the gravitational field of the earth into a parabola. Its curvature depends only on the velocity v of the particles, not on their mass m. In order to determine the momentum $p=mv$ one has to find out the absolute value of m. For this purpose one may use the Perrin method. A rarefied "gas" of the matter is made and the decrease of its concentration ρ with the altitude z is observed. Brownian particles of the known mass M are then suspended in the gas and the corresponding decrease of their number N with the altitude is counted. If the temperature T is not too low, one has, as the result of mechanical collisions, the Maxwell-Boltzmann ratios*

$$\frac{N_1}{N_2}=e^{-Mg(z_1-z_2)/kT} \quad \text{and} \quad \frac{\rho_1}{\rho_2}=e^{-mg(z_1-z_2)/kT}$$

at two altitudes z_1 and z_2, leading to the mass ratio

$$\frac{m}{M}=\frac{\log(\rho_1/\rho_2)}{\log(N_1/N_2)}.$$

Thus a comparison is obtained between the characteristic molecular weight m to be assigned to the matter and the known weight M of the Brownian particles. With the help of m one knows also the value of the momentum $p=mv$ of the particles which are supposed to constitute the parabolic beam of matter.

If we suppose instead that the beam consists of *waves* of a constant frequency ν, we can explain its parabolic form on the hypothesis that the earth produces a refractive index $n_\nu(z)$ for waves varying with the altitude. In order to determine the *absolute* value of the wave length λ we may send the beam onto a body with a large gradient in its index of refraction, for instance onto an artificial grating with a known distance d between successive grooves. A natural crystal lattice will serve the same purpose if its grating constant d is known. The resulting diffraction pattern with its maxima and minima of intensity determines then the inherent wave length λ of the matter beam according to the wave theory.

The absolute values of λ and p found in these ways will turn out to possess the universal constant product

$$p\,.\,\lambda=h=6{\cdot}55\times10^{-27}\,\text{gr. cm.}^2\,\text{sec.}^{-1}=\text{Planck's constant.}$$

This is the formula of de Broglie. Planck's quantum of action h appears as the link between the two classical theories of matter.

* Instead of saying a body has the absolute temperature T one can say that a free particle in thermal equilibrium with the body has the average energy $E=\frac{3}{2}kT$. Thus T is only another energy scale besides the erg-scale, and k relates the two arbitrary scales of energy. It is not true that k is another universal constant like e, m, c, h.

§8. DOUBLE RAY OF MATTER DIFFRACTING LIGHT WAVES

We are now going to discuss a more complicated standard experiment which still can be explained with the help of either classical theory. This example will lead us, however, to a more intimate knowledge of the rules of quantum theory which form the link between corpuscular and wave data. A homogeneous beam of matter whose corpuscular momentum p and wave length λ have been determined according to §7 is allowed to be reflected from a wall, so that it returns along the same path in the $-x$-direction, forming what we may call a *double ray* of matter, or, from the wave point of view, a *linear gas crystal*. In order to get information about its qualities we illuminate the double ray with monochromatic light. It will then happen that the double ray of matter serves to split up the incident light without change of colour into *two* directions $+\alpha$ and $-\alpha$ (Fig. 2).* This "coherent diffraction", that is, deflection without change of colour, can be interpreted in two ways.

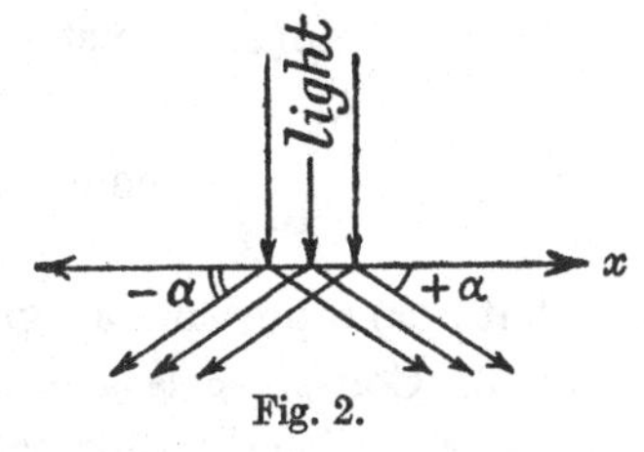

Fig. 2.

According to the *wave theory* the various line elements dx of the matter ray serve as Huygens centres of secondary light waves. If $\rho(x)$ is the material density of the double ray along the x-axis, and if light of the wave length Λ is incident from a perpendicular direction y, we expect according to the wave theory to observe in the direction of α the superposed light amplitude

$$(5) \qquad A(\alpha) = \text{const.} \int \rho(x) \cos\left[2\pi\nu t + \frac{2\pi}{\Lambda} x \cos\alpha\right] dx.$$

Here $x\cos\alpha$ is the path difference of the secondary light ray emerging from the point x towards the direction of α, as compared

* The greater part of the incident light intensity will be scattered incoherently with change of colour. We consider here only the coherent diffraction (p. 17).

with the path of the ray emerging from $x=0$ (Fig. 2). Using the complex form

$$A(\alpha)=\text{const.}\, e^{2i\pi\nu t}\int \rho(x)\, e^{\frac{2i\pi}{\Lambda}x\cos\alpha}\,dx, \tag{5'}$$

we obtain the light intensity observed in the direction of α to be the absolute value $I(\alpha)=|A(\alpha)|^2$.

In order to explain that $A(\alpha)$ is 0 except for two selected directions $\pm\alpha$, we have to assume that $\rho(x)$ has the form of a *periodic density*

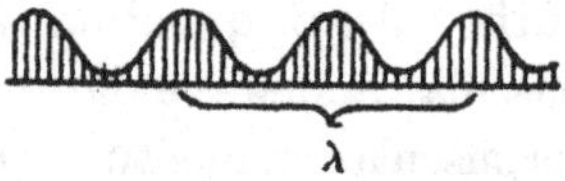

λ

Fig. 3.

$$\rho(x)=\text{const.}+\bar{\rho}\,.\,2\cos^2\left(\frac{2\pi}{\lambda}x+\phi\right) \tag{6}$$

$$=(\text{const.}+\bar{\rho})+\bar{\rho}\,.\cos\left(2\pi x\frac{2}{\lambda}+2\phi\right)$$

so that $\rho(x)$ represents a "grating" with the grating constant $\lambda/2$ (Fig. 3). Const. $+\bar{\rho}$ is the average density, and the phase ϕ in the cos-function depends on the choice of the zero point of the x-axis. Indeed, if (6) is inserted for ρ in (5'), the integral (5') vanishes unless the periodicity of ρ is equal to that of the exponential function under the integral (5'), that is unless the condition

$$\frac{\cos\alpha}{\Lambda}=\frac{2}{\lambda} \quad \text{or} \quad \frac{\lambda}{2}\cos\alpha=\Lambda \tag{6'}$$

is satisfied. This is a diffraction of the first order. Conversely, if the angle α of the deflected light is measured, one may find the hypothetical grating constant $\lambda/2$ of the double ray from the formula (6').

The diffracted intensity $I=|A|^2$ increases proportionally with $(\bar{\rho})^2$. On the other hand, if only a finite interval X of the double ray containing $X\,.\,2/\lambda$ density maxima is illuminated, then the evaluation of (5') shows that the peak of the intensity maximum at α increases, as the square X^2 of the length of the grating. On the other hand, the width of the maximum decreases as $1/X$, so that the total diffracted intensity increases proportionally with the length X. This is in agreement with the elementary theory of gratings.

There is no reason for interpreting $\rho(x)$ as a strictly continuous distribution of matter along the x-axis. The same diffraction effect would result if $\rho(x)$ described only a *statistical distribution* of particles along the x-axis. Such a corpuscular interpretation of ρ would conflict, however, with the rules of dynamics. For instance, it would be impossible to reconcile the apparent absence of matter in the nodes of $\rho(x)$ with the idea of two corpuscular matter rays travelling along $\pm x$. But we have to remember that the density function $\rho(x)$ resulted from our using an hypothesis—the wave theory of light. The derived quantity ρ contains the feature of this hypothesis in its turn. If we had used the corpuscular theory of light for explaining the light diffraction towards $\pm\alpha$, then we should have come to quite a different image of the structure of the ray of matter (§ 9) in which there is no indication of density maxima and minima at all.

In addition to the "coherent diffraction" without change of colour at angles $\pm\alpha$ there will be a scattering of the incident parallel light into other directions α', connected, however, with a change of colour. This incoherent scattering, if interpreted according to the wave theory, would appear to have its Huygens source in various "transition densities" ρ'. In particular, the scattered light of the wave length Λ' observed in the direction of α' appears to be sent out from Λ'-emitting synchronous sources distributed along the x-axis with the transition density

$$\rho'(x) = \text{const.} + \bar{\rho}' \,.\, 2\cos^2\left(\frac{2\pi x}{\lambda'} + \phi'\right),$$

where λ' is determined by $\dfrac{\cos\alpha'}{\Lambda'} = \dfrac{2}{\lambda'}$ in analogy to (6), (6').

§ 9. DOUBLE RAY OF MATTER DIFFRACTING PHOTONS

Let us now attempt to obtain information about the physical properties of the double ray of matter by interpreting the optical diffraction pattern of Fig. 2 by means of the *corpuscular theory.*

Instead of light waves of wave length Λ we now suppose that photons possessing the momentum $P = h/\Lambda$ are deflected at angles $\pm\alpha$, without change of their energy and their total momentum. The x-component of P was zero before and is $\pm P.\cos\alpha$ after the deflection. This increase must correspond to an equal decrease of the x-momentum of the matter ray. The y-component of P might be allowed to change without compensation if we consider the matter particles to be lined up on a rigid x-axis in our linear example. So we are led to consider the double ray as consisting of two groups of particles of matter, one carrying the x-momentum $+p+$constant and the other group carrying $-p+$the same constant, the constant having no physical significance. The deflection of a photon would be due to an impact in which a particle of matter changes over from the $+p$ group to the $-p$ group or vice versa. The total momentum of the matter ray would then be increased in each impact by $\pm 2p$ according to the equation

(7) $P.\cos\alpha = \pm 2p$ (conservation of momentum).

This equation is the corpuscular equivalent of the wave relation (6′). Since P corresponds to h/Λ, we obtain equivalence of (6′) and (7′) if we make p correspond to h/λ.

It is not necessary to explain the conservation formula $P.\cos\alpha = 2p$ as meaning that just *one* particle of matter jumps from $+p$ to $-p$ during the interaction with light. One could say just as reasonably that $N+1$ particles of the matter ray jump from $+p$ to $-p$ and simultaneously N particles from $-p$ to $+p$. The latter assumption would comply better with the intensity rules of the coherent diffraction of light (refer to § 15).

In general, a linear matter ray contains particles of various amounts $p', p'', \ldots$ of momentum with *abundances* $\sigma(p')$, $\sigma(p'')$, ... per unit of length. In the particular case of our double ray we have the abundance function

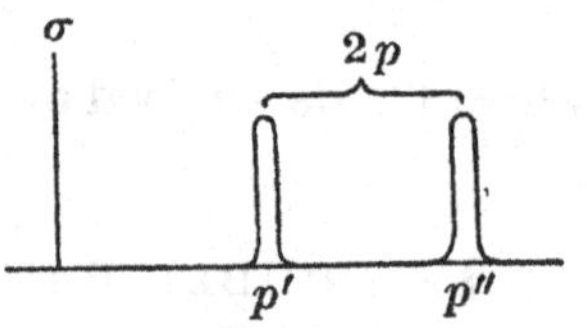

Fig. 4.

(8) $\sigma(p) \equiv 0$ except for $p = p'$ and $p = p''$,

where $p' - p'' = 2p$ and $\sigma(p') = \sigma(p'') = \bar{\rho}/2$, cf. Fig. 4.

This abundance function indicates that the double ray is bound to give off and take on momenta

$$p'-p''=2p=2\frac{h}{\lambda}$$

without being transformed into another physical state. The absolute values p' and p'' (zero point of the p-scale) cannot be determined by optical observation. This corresponds to the impossibility of observing the phase ϕ of the periodic function $\rho(x)$.

The incoherent scattering of photons into directions other than $\pm\alpha$ means that the matter ray at other times gives off or takes on momenta other than $2p=2h/\lambda$, so that the double ray is broken up and transformed into a "multiple ray" of matter containing momenta other than p' and p''.

If a homogeneous *single ray* is illuminated with parallel light of wave length Λ, one observes only an incoherent diffuse scattering but no coherent diffraction at all. According to the wave theory of light this means that the density function $\rho(x)$ of the single ray is a constant. The corpuscular theory of light tells us that the incident photons get either no impulses from the matter ray or only impulses which transform the matter ray into another state. Hence the original matter ray can have only *one* characteristic momentum p', no changes to any other p'' being possible within its scope. Its abundance function is

$$\sigma(p)=0 \text{ except for a certain } p=p'.$$

§ 10. MICROSCOPIC OBSERVATION OF $\rho(x)$ AND $\sigma(p)$

One may ask whether a microscopic observation of a double ray of matter may show more directly the existence of those maxima and minima of $\rho(x)$. In order to decide this question one may illuminate through a narrow slit of the width $\Delta x<\lambda/2$ a small section Δx of the matter ray with parallel light of the wave length Λ. If the slit is opposite a maximum of $\rho(x)$, then one would expect that the incident light Λ will be diffracted by the illuminated matter, the maximum of matter becoming visible in this way. If the slit is just opposite a minimum of $\rho(x)$, then no such diffraction of the light

Λ should occur. Since a maximum of ρ covers a length $\lambda/2$ of the x-axis, it would produce a coherent diffraction over an angular range $\pm\alpha'$ determined by $\cos\alpha' = 2\Lambda/\lambda$. But the narrow slit $\Delta x < \lambda/2$ itself diffracts the light Λ over the wider range $\cos\alpha'' > 2\Lambda/\lambda$, no matter whether the slit is before a maximum or before a minimum of ρ. Thus the absolute position of the individual nodes and loops of $\rho(x)$ cannot be located. All this applies only to the coherent diffraction at the two selected directions $\pm\alpha$ determined by $\cos\alpha = 2\Lambda/\lambda$ (6′), the incoherent scattering being linked with a quite different transition density $\rho' \neq \rho$.

$\rho(x)$ cannot be told from $\rho(x + \text{const.})$. On the other hand, since $\sigma(p)$ is derived only from observing transitions of momentum, $\sigma(p)$ cannot be told from $\sigma(p + \text{const.})$.

The case of a ray of free particles reflected from a wall (standing waves of matter in a medium of constant index of refraction) is quite different from a ray travelling in a fixed periodic field of force originating from equidistant lattice points (periodic index of refraction for waves). Here not only ρ but also all the transition densities ρ' would show the periodicity of the lattice, and a microscopic observation of density maxima and minima would be possible here. Our case of a double ray formed by free particles reflected from a wall may be termed a *linear gas crystal*; the latter case of a permanent lattice represents then a *solid crystal*.

§ 11. COMPLEMENTARITY

Let us now sum up the result of the preceding considerations. The optical observation of the matter ray leads on the one hand to the *density function*

$$\text{(6)} \qquad \rho(x) = \text{const.} + \bar{\rho} \,.\, 2\cos^2\left(\frac{2\pi}{\lambda}x + \phi\right)$$

$$= (\text{const.} + \bar{\rho}) + \bar{\rho} \,.\, \cos\left(2\pi x \frac{2}{\lambda} + 2\phi\right)$$

as the Huygens source of secondary light waves. On the other hand, writing p for p_x, we were led to the *abundance function*:

$$\text{(8)} \qquad \sigma(p) \equiv 0 \text{ except for } p' = \frac{h}{\lambda} + \text{const. and } p'' = -\frac{h}{\lambda} + \text{const.}$$

$$(8') \qquad \text{with } \sigma(p') = \sigma(p'') = \frac{\bar{\rho}}{2} \quad \text{and} \quad p' - p'' = \frac{2h}{\lambda},$$

determining the momentum $(p'-p'')$ which the ray can give off, as judged from the corpuscular point of view. The two functions $\rho(x)$ and $\sigma(p)$ are inseparably bound together as *complementary properties* of the double ray. Both $\rho(x)$ and $\sigma(p)$ describe the same activity of the ray, its ability to diffract incident monochromatic light through two selected directions $\pm\alpha$ without change of colour.

We are confronted here with a significant feature of matter (and of light), the correlation of two simultaneous properties $\rho(x)$ and $\sigma(p)$ of one and the same object, *both properties accounting for the same physical activity in two different interpretations.*

The main purpose of quantum theory is to find a direct formal *mathematical* relation between two such complementary properties as $\rho(x)$ and $\sigma(p)$, without resorting in every instance to a discussion of optical observations.

If such mathematical relations exist, so that $\rho(x)$ determines $\sigma(p)$, and $\sigma(p)$ determines $\rho(x)$, then we can foresee a significant result of quantum theory. Since both $\rho(x)$ and $\sigma(p)$ represent *physical* functions satisfying certain natural conditions of uniqueness and finiteness, only such functions $\sigma(p)$ are admissible that correspond to *physical* (unique and finite) functions $\rho(x)$. Likewise not every physical function $\sigma(p)$ can be admitted, but only such functions σ that correspond to *physical* functions $\rho(x)$. In this way we sometimes are compelled to confine the choice of the mutually dependent functions (as ρ or σ) to a certain selection; and sometimes this selection will consist of a discontinuous set of possible physical states of a material system. In this way Schrödinger explained the existence of *quantized states* of the corpuscular energy E as complementary to eigen-vibrations of frequency $\nu = E/h$.

§ 12. MATHEMATICAL RELATION BETWEEN $\rho(x)$ AND $\sigma(p)$ FOR FREE PARTICLES

There is a direct mathematical relation between the observable density function $\rho(x)$ of (6) and the observable abundance function $\sigma(p)$ of (8), not involving direct reference to an optical observation.

Let us start, for instance, with the abundance function $\sigma(p)$ of (8) which vanishes except for the two arguments p' and p''. Then define the complex "abundance amplitudes" $\chi(p')$ and $\chi(p'')$:

$$\chi(p')=\sqrt{\sigma(p')}e^{i\delta'} \quad \text{and} \quad \chi(p'')=\sqrt{\sigma(p'')}e^{i\delta''}, \tag{9}$$

whose absolute squares are $\sigma(p')$ and $\sigma(p'')$, but which still contain certain phases δ' and δ'' to be fixed later. Now form two complex "wave functions" using $\chi(p')$ and $\chi(p'')$ as their amplitudes:

$$\chi(p')e^{2\pi i\frac{p'}{h}x} \quad \text{and} \quad \chi(p'')e^{2\pi i\frac{p''}{h}x}. \tag{9'}$$

Their wave lengths are $\lambda'=h/p'$ and $\lambda''=h/p''$. Superpose them and get the resulting wave amplitude

$$\psi(x)=\chi(p')e^{2\pi i\frac{p'}{h}x}+\chi(p'')e^{2\pi i\frac{p''}{h}x}. \tag{10}$$

Finally define $\rho(x)$ as the absolute square† (intensity) of $\psi(x)$:

$$\begin{aligned} \rho(x) &= |\psi(x)|^2=\psi(x).\psi^*(x) \\ &= |\chi(p')|^2+|\chi(p'')|^2 \\ &\quad +\chi(p')\chi^*(p'')e^{\frac{2\pi i}{h}(p'-p'')x}+\chi^*(p')\chi(p'')e^{-\frac{2\pi i}{h}(p'-p'')x} \\ &= \sigma(p')+\sigma(p'')+ \\ &\quad \sqrt{\sigma(p')\sigma(p'')}\,2\cos\left[\frac{2\pi}{h}(p'-p'')x+(\delta'-\delta'')\right]. \end{aligned} \tag{10'}$$

The result is seen to be identical with (6) when account is taken of (8′). In conclusion: Starting with an abundance function $\sigma(p)$, one forms first an "abundance amplitude"

$$\chi(p)=\sqrt{\sigma(p)}e^{i\delta(p)}$$

containing an indefinite phase $\delta(p)$. Then one builds up the "density amplitude" $\psi(x)$ according to (10) and finally one defines $\rho(x)$ as the absolute square of $\psi(x)$. The relation between $\sigma(p)$ and $\rho(x)$ is given by the scheme

$$\sigma(p)\rightarrow\chi(p)\rightarrow\psi(x)\rightarrow\rho(x).$$

The reverse process is just as feasible. If both σ and ρ could be measured (see however § 10), then the phases δ would all be determined up to an additive constant.

† The asterisk stands for complex conjugate.

The intermediary complex density amplitude $\psi(x)$ is often called the "probability amplitude". This term is derived from the idea that the density $\rho(x)$ describes a statistical distribution of corpuscles along x so that $\rho(x)dx$ measures the probability of a particle being found just between x and $x+dx$. (See, however, § 2 (*e*).)

(10′) expresses the so-called *interference of probabilities*. That is, if there were only *one* homogeneous matter ray with particles of momentum p' and present with the abundance $\sigma(p')$ per unit of length, one would expect the probability of finding one of these particles per unit of length x to be $\rho'(x)=\sigma(p')$, and $\rho'(x)$ would be constant along x. If another beam of particles of momentum p'' were present with the abundance $\sigma(p'')$, one would expect a probability $\rho''(x)=\sigma(p'')=$ const. If both matter rays were present simultaneously, one would expect the probability of finding a particle per unit of length to be the sum

$$\rho'+\rho''=\sigma(p')+\sigma(p'').$$

Instead (10′) tells us that the density along a double ray, that is, the probability for one particle to be found along a unit of length, is given by the sum of the two constant terms plus an "interference term". The origin of this extra term can be described thus: The "wave intensity" $\rho(x)$, instead of being the *sum of the absolute squares* of the two "wave amplitudes" (9′), is the *absolute square* (10′) *of their sum*, analogous to the explanation of interference in wave optics.

There are however grave objections to this corpuscular interpretation of the density function $\rho(x)$. If we receive a message reading "bridge", we can interpret it in two independent ways. We may suppose it either to come from people on a bridge, or from people playing the game of bridge. Either theory explains the message completely. But it would be unreasonable to infer that the message is sent out by people playing bridge on a bridge. In the case of our double ray we have to interpret the message of the optical diffraction pattern $\pm\alpha$. Either it is due to a periodic density $\rho(x)$ (wave theory) or to particles $\pm p$ that collide with photons (corpuscular theory). But it would be unreasonable to assume that the pattern comes from particles $\pm p$ that are distributed like the periodic density $\rho(x)$.

A similar over-interpretation would be made if we should establish a *wave interpretation of the corpuscular transitions* $+p \rightarrow -p$, saying that the deflections of the photons are due to mechanical impulses given out not by the particles $\pm p$ but by the standing wave $\rho(x)$. In either case we should be guilty of violating the very idea of quantum theory, that the corpuscular and the wave theory are independent pictures which cannot be fused.† In contrast to the two quite independent explanations of the "bridge" message, however, in quantum theory the two apparently independent observables, the density distribution $\rho(x)$ and the abundance distribution $\sigma(p)$, are intimately connected by formal mathematical relations which express the physical fact that both ρ and σ spring from different interpretations of the same optical observations. It is only if one insists on clinging to the old mechanical models that one comes to such contradictory ideas as "interference of probabilities", "corpuscles guided by wave rules", "vibrations with quantised energy", "failure of mechanical causality" and the like.

The physical meaning of $\rho(x)$ is much better expressed if we call it the "transition density". That is, $\rho(x)$ is the wave density which we need for explaining the diffraction pattern $\pm\alpha$ in wave terms, corresponding to transitions $+p \rightarrow -p$ if the same pattern

† An objection could be found in the fact that the density function of the wave theory often allows us to predict the abundance of particles appearing in various regions of space. Such cases, however, concern measurements in macroscopic dimensions; and we have just shown in the previous sections that the observed macroscopic intensity distribution in such a pattern can be explained by *both* theories—as far as time averages are concerned. Remembering what we learned at the end of § 3 about fluctuations, we see that the wave theory can be used for calculating the average intensity distribution of a diffraction pattern, and yet corpuscular fluctuations will be observed in the case of a small absolute intensity. On the other hand, we can employ just as well the corpuscular theory for calculating the average intensity distribution and yet find interference fluctuations in the case of a large absolute intensity.

is interpreted in corpuscular language. In order to indicate that the wave density *corresponds* to a complementary corpuscular transition $p' \to p''$ one usually writes $\rho_{p'p''}(x)$. In the case of a stationary state (transition of p into itself) one might write $\rho_{pp}(x)$ or $\rho_p(x)$.

§ 13. GENERAL RELATION BETWEEN $\rho(q)$ AND $\sigma(p)$

The method developed for the double ray can be easily generalized for a "multiple ray" consisting of a number of components $p', p'', \ldots$ present with the abundances (per unit of length on the q-axis)

$$\sigma(p') = \sigma', \quad \sigma(p'') = \sigma'', \ \ldots.$$

In order to obtain the corresponding density function $\rho(x)$ we must first form the "abundance amplitudes"

$$\chi(p') = \sqrt{\sigma'} e^{i\delta'}, \quad \chi(p'') = \sqrt{\sigma''} e^{i\delta''}, \ \ldots \tag{11}$$

containing indefinite phases δ, only the differences of which appear in the final expression for ρ. We then form the "density amplitude" along the q-axis as the sum

$$\psi(q) = \Sigma \chi(p') e^{\frac{2\pi i}{h} qp'} = \Sigma \sqrt{\sigma'}\, e^{\frac{2\pi i}{h} qp' + i\delta'}, \tag{12}$$

each amplitude $\chi(p')$ being multiplied by a complex wave function of wave length $\lambda' = h/p'$. Finally, we define the density by

$$\rho(q) = |\psi(q)|^2 = \psi(q)\psi^*(q),$$

that is,

$$\rho(q) = \Sigma\sigma' + \Sigma\Sigma \sqrt{\sigma'\sigma''}\, 2\cos\left[\frac{2\pi}{h} q(p' - p'') + (\delta' - \delta'')\right] \tag{12'}$$

Due to the indefinite phases $\delta', \delta'', \ldots$ it is not possible to predict, uniquely, the density function $\rho(q)$ from the given abundance $\sigma(p)$ and vice versa.†

† We shall see later (§ 18) that the δ' are linear functions of the time $\delta' = 2\pi\nu' t + \gamma'$, where $\nu' = E'/h$. Jumps without change of the energy (like $+p \to -p$) lead then to sum terms in (12′) that are constant in time (standing waves); whereas energy changes $E' - E''$ contribute running beats of frequency $\nu' - \nu''$.

The inverse process of finding $\sigma(p)$ from $\rho(q)$ runs as follows. First, introduce the density amplitude

$$\psi(q) = \sqrt{\rho(q)}\, e^{i\phi(q)}, \tag{13}$$

where $\phi(q)$ is an undetermined phase function. Suppose, now, that the multiple ray extends only along a finite though very large length Q. Or if the ray is infinitely long, the values of $\rho(q)$ may repeat after a sufficiently long interval Q of q, so that $\rho(q)$ has the periodicity Q. Now define the abundance amplitude $\chi(p)$ per unit of length as the integral over Q:

$$\chi(p) = \frac{1}{Q}\int \psi(q)\, e^{-\frac{2\pi i}{h}p.q}\, dq = \frac{1}{Q}\int \sqrt{\rho(q)}\, e^{-\frac{2\pi i}{h}p.q + i\phi(q)}. \tag{14}$$

This equation is the inverse of equation (12); indeed, if the series (12) for $\psi(q)$ is inserted into (14), the integral becomes zero for all values p except for $p = p', p'', p''', \ldots$, where it takes on the values $\chi(p'), \chi(p''), \ldots$, provided that Q is so long as to contain many wave lengths $\lambda', \lambda'', \lambda''', \ldots$. Finally, define $\rho(p)$ as the absolute square of $\chi(p)$:

$$\sigma(p) = |\chi(p)|^2 = \chi(p)\chi^*(p) \tag{14'}$$

$$= \frac{1}{Q^2}\iint \sqrt{\rho(q')}.\sqrt{\rho(q'')}\, e^{-\frac{2\pi i}{h}p(q'-q'') + i(\phi'-\phi'')}\, dq'dq''.$$

The abundance $\sigma(p)$ cannot be predicted uniquely from the density $\rho(q)$ because of the undetermined phase function $\phi(q)$.

In conclusion, the relation between the observables $\sigma(p)$ and $\rho(q)$ is contained in the relation between their "amplitudes":

$$(15)\quad \begin{cases} (a)\ \ \psi(q) = \Sigma\chi(p')\, e^{\frac{2\pi i}{h}p'.q} \quad \text{and} \quad |\psi|^2 = \rho, \\ (b)\ \ \chi(p) = \dfrac{1}{Q}\displaystyle\int \psi(q)\, e^{-\frac{2\pi i}{h}q.p}\, dq \quad \text{and} \quad |\chi|^2 = \sigma. \end{cases}$$

(15*b*) is the mathematical inverse of (15*a*). If $\sigma(p)$ is a *continuous* *function* of p in which *all* values of p (not only $p', p'', \ldots$) are represented in the multiple ray, the following integral formulae result:

$$(16)\begin{cases}(a)\quad \psi(q)=\int_{-\infty}^{\infty}\chi(p)e^{\frac{2\pi i}{h}p.q}dp \quad\text{and}\quad |\psi|^2=\rho,\\ (b)\quad \chi(p)=\frac{1}{h}\int_{-\infty}^{\infty}\psi(q)e^{-\frac{2\pi i}{h}p.q}dq \quad\text{and}\quad |\chi|^2=\sigma.\end{cases}$$

(16b) is the mathematical inverse of (16a).

Thus far we have considered matter rays along a linear axis q and with momenta p parallel to this axis. All results can immediately be generalized for *currents of matter in space*. The density function in various points of space is now $\rho(q)=\rho(x,y,z)$, q representing a radius vector with the three components x, y, z. The abundance of various vectorial momenta p with the components p_x, p_y, p_z is described by an abundance function $\sigma(p)=\sigma(p_x, p_y, p_z)$. The relation between ρ and σ will again be controlled by intermediate amplitude functions $\chi(p_x, p_y, p_z)$ and $\psi(x,y,z)$, which are connected by relations like (15), (16), if in all these formulae p and q represent *vectors*, and products $p.q$ represent the scalar products

$$(16')\qquad p.q=p_x.x+p_y.y+p_z.z.$$

The relation between $\psi(q)$ and $\chi(p)$ is that of a *Fourier expansion* with respect to the periodic functions $e^{\pm\frac{2\pi i}{h}pq}$. In (15$a$) the $\chi(p')$ are the coefficients of the expansion of $\psi(q)$ into a series of periodic functions. And in (15b) $\psi(q)$ is the coefficient in an expansion of $\chi(p)$ into a Fourier integral. This relationship between the density amplitude and the abundance amplitude as expansions of one another with respect to periodic functions is the basic theorem of quantum mechanics for free particles. It originated from the fact that $\rho(q)$ was derived from optical observations of the matter as interpreted by light waves Λ, while $\sigma(p)$ was derived from the same observations as interpreted by photons of the momentum P, where $P.\Lambda=h$.

§ 14. CRYSTALS

Of particular interest is the case of a system of matter currents $p', p'', \ldots$ whose components are positive or negative integral multiples of certain fundamental values p_x^0, p_y^0, p_z^0. For instance, p' may have the components

$$(17) \qquad p_x' = k'.p_x^0, \quad p_y' = l'.p_y^0, \quad p_z' = m'.p_z^0,$$

where $k' \, l' \, m'$ is a triplet of integers. At the same time p'' may be characterized by another triplet of integers $k'' \, l'' \, m''$. If $\sigma', \sigma'', \ldots$ are the abundances of the various momenta, then the density amplitude in space according to (12) is

$$\psi(q) = \Sigma \chi(p') e^{\frac{2\pi i}{h}(q.p')}, \quad \text{where} \quad q.p' = x.k'p_x^0 + y.l'p_y^0 + z.m'p_z^0,$$

the sum extending over all sets of the integers k', l', m' from $-\infty$ to $+\infty$. The χ's here are related to the σ's in (11). The density $\rho(x,y,z) = \rho(q) = |\psi(q)|^2$ then becomes, according to (12'),

$$(18) \quad \rho(q) = \sigma' + \sigma'' + 2\sqrt{\sigma'\sigma''} \cos\left[\frac{2\pi}{h}(q.p'-p'') + (\delta' - \delta'')\right] + \ldots,$$

where

$$(q.p'-p'') = x.p_x^0(k'-k'') + y.p_y^0(l'-l'') + z.p_z^0(m'-m'').$$

$\rho(q)$ is a Fourier series representing a periodic function with the three periodicities

$$(17') \quad \lambda_1 = \frac{h}{p_x^0}, \quad \lambda_2 = \frac{h}{p_y^0}, \quad \lambda_3 = \frac{h}{p_z^0} \quad \text{in } x, y, z \text{ respectively.}$$

The density distribution is that of a rectangular *crystal* with the three grating constants $\lambda_1, \lambda_2, \lambda_3$.

Thus we have found two quite equivalent definitions of a crystal. First, it is a periodic distribution of matter with the periodicities $\lambda_1, \lambda_2, \lambda_3$ along three directions x, y, z. Second, it is an assembly of matter currents that give out momenta p_x, p_y, p_z which are integral multiples of certain basic momenta p_x^0, p_y^0, p_z^0. These two descriptions are equivalent and inseparably coupled together with no preference to be given to either of them. This equivalence can be illustrated now if we describe the reaction of a crystal to incident light (X-rays).

M. Laue's theory of X-ray diffraction in crystals tells us that waves Λ, incident in a direction α_i, β_i, γ_i, are diffracted into such selected directions α_j, β_j, γ_j (Fig. 5) that the following three equations are satisfied simultaneously:

$$(19)\quad \begin{cases} \lambda_1(\cos\alpha_i-\cos\alpha_j)=k\Lambda, \\ \lambda_2(\cos\beta_i-\cos\beta_j)=l\Lambda, \\ \lambda_3(\cos\gamma_i-\cos\gamma_j)=m\Lambda. \end{cases} \quad \text{(Laue's interference rules)}$$

The Laue spots represent "needles" of radiation resulting from interference.

It was first realized by W. Duane (8) that these diffraction rules are equivalent to the mechanical conservation rules for the three components of the total momentum during a collision between the crystal and a photon P. The x-component of the momentum of the photon changes by

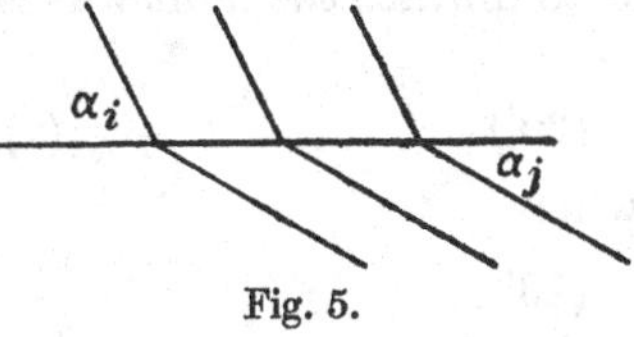

Fig. 5.

$$P_x^i-P_x^j=P(\cos\alpha_i-\cos\alpha_j).$$

This must be equal to the simultaneous increase of the x-momentum of the crystal:

$$p_x''-p_x'=(k''-k')p_x^0=kp_x^0$$

and similarly for the y- and z-components. So one obtains the equations

$$(19')\quad \begin{cases} P(\cos\alpha_i-\cos\alpha_j)=kp_x^0, \\ P(\cos\beta_i-\cos\beta_j)=lp_y^0, \\ P(\cos\gamma_i-\cos\gamma_j)=mp_z^0. \end{cases} \quad \text{(Conservation of momentum)}$$

Owing to the large mass of the crystal, the energy of the latter and the energy of the photon are not changed during a collision. The two rules (19) and (19′) prove to be identical on account of the relations (17′) between corpuscular and wave quantities. This twofold discussion of the optical experiment verifies the mathematical relations (12), (14) of quantum theory.

§15. TRANSITION DENSITY AND TRANSITION PROBABILITY

Let us again consider a long linear matter ray containing various momenta p. The matter ray may be illuminated from the direction of α_0 and the diffracted light may be observed in the direction α. As observed from this angle, the matter ray appears *either* to give out momenta $(p'-p'')$ to the photons P determined by

$$(20) \quad p'-p''=P(\cos\alpha-\cos\alpha_0) \quad \text{(Conservation of momentum)}$$

or to diffract the light waves Λ by means of a density distribution

$$(20') \qquad \rho_{\alpha_0\alpha}(x)=\text{const.}\, e^{-\frac{2\pi i}{\lambda}x},$$

where

$$(20'') \qquad \frac{1}{\lambda}=\frac{\cos\alpha-\cos\alpha_0}{\Lambda}.$$

(Indeed, when we insert (20′) in the formula

$$A(\alpha)=\text{const.}\int\rho(x)\,.\,e^{\frac{2\pi i}{\Lambda}(\cos\alpha-\cos\alpha_0)x}\,dx$$

of the diffracted amplitude, we find $A(\alpha)\equiv 0$ unless the integrand in

$$A(\alpha)=\text{const.}\int e^{2\pi i\left(-\frac{1}{\lambda}+\frac{(\cos\alpha-\cos\alpha_0)}{\Lambda}\right)x}\,dx$$

is independent of x, that is unless (20″) is satisfied.) The corpuscular relation (20) and the wave relation (20″) prove to be identical on account of the quantum formulae

$$(21) \qquad P=\frac{h}{\Lambda} \text{ and } \frac{1}{\lambda}=\frac{p'-p''}{h}.$$

The density $\rho_{\alpha_0\alpha}(x)$, which can be considered responsible for the deflection $\alpha_0\rightarrow\alpha$, corresponds to the corpuscular transition $p'\rightarrow p''$, and this can be expressed by changing the subscripts $\alpha_0\alpha$ to $p'p''$:

$$(22) \quad \rho_{\alpha_0\alpha}(x)=\text{const.}\, e^{-\frac{2\pi i}{\Lambda}(\cos\alpha-\cos\alpha_0)x} = \rho_{p'p''}(x)=\text{const.}\, e^{-\frac{2\pi i}{h}(p'-p'')x}.$$

As seen from a particular angle α, and interpreted according to the wave theory, the matter ray displays a particular "transition density" $\rho_{\alpha_0\alpha}=\rho_{p'p''}$.

The density function

$$\rho(x) = \text{const.} + \bar{\rho}\left[1 + \cos\left(\frac{2\pi}{h}(p' - p'')x\right)\right]$$

of the double ray is the sum of the two transition densities $\rho_{p'p''}$ and $\rho_{p''p'}$ of (22) plus a constant, the latter having no physical significance. Comparison of (22) with (10′) shows that the constant amplitude of $\rho_{p'p''}$ is the product $\chi(p') . \chi^*(p'')$. Thus we obtain the formula for the transition density:

$$\rho_{p'p''}(x) = \chi(p') . \chi^*(p'') . e^{\frac{2\pi i}{h}(p'-p'')x}. \tag{23}$$

The deflected intensity $I(\alpha) = |A(\alpha)|^2$ becomes then

$$\begin{aligned} I(\alpha) &= \text{const.}\, |\chi(p') . \chi^*(p'')|^2 \\ &= \text{const.}\, \sigma(p') . \sigma(p'') = \text{const.}\, N' . N'', \end{aligned} \tag{24}$$

where N' and N'' are the numbers of particles of matter with momenta p' and p'' present in the matter ray. This reads in corpuscular language that the transition processes $p' \to p''$ which are responsible for the deflected intensity $I(\alpha)$ occur with a probability

$$\begin{aligned} &\text{const.}\, \sigma(p') . \sigma(p'') = \text{const.}\, N' . N'' \\ &= \text{transition probability} = \text{const.}\, I(\alpha). \end{aligned} \tag{24'}$$

[to be corrected in (25)]

$I(\alpha)$ increases by a factor of N^2 if the $\sigma(p')$ and $\sigma(p'')$, and thus the total abundance of matter, are increased by a factor of N (for instance by taking a matter ray N times as long or N times as dense). This corresponds to the well-known fact that the altitude of the intensity maximum produced by a grating increases as the square of the number N of interfering light rays. But since on the other hand its width decreases by the same factor, the total intensity radiated into this secondary maximum will increase only proportionally to N itself. The same will be true for the diffraction effect of our line of matter. True, the intensity radiated into the peak of the maximum $I(\alpha)$ will be proportional to $\sigma(p') . \sigma(p'')$ or to $N' . N''$; and this number may be interpreted in corpuscular language as the *excess number* of transitions $p' \to p''$ over the transitions $p'' \to p'$. But the intensity of the whole width of the light maximum will be proportional only to $\sqrt{\sigma(p') . \sigma(p'')}$ or to

$\sqrt{N'.N''}$, that is, proportional to the geometric mean of the abundances of matter in the initial and final states of the transition.

A more elaborate theory which applies quantum theory not only to the matter but also to the light shows, however, that a correction to the foregoing formulae must be introduced. According to (24), no diffracted intensity at all would be observed at the angle α if one of the abundance functions $\sigma(p')$ or $\sigma(p'')$ belonging to the transition

$$p' - p'' = \frac{h}{\Lambda}(\cos\alpha - \cos\alpha_0)$$

were *zero*. In reality, transitions $p' \to p''$ happen even if $\sigma(p'')$ is vanishingly small. The corrected theory of radiation of Einstein and Dirac(9) shows that (24) should be replaced by

$$(25) \qquad I(\alpha) = \text{const.}\, N'(N''+1),$$

so that in the special case of $N''=0$, $I(\alpha)$ is still proportional to N' or to $\sigma(p')$ in the initial state. A corresponding correction is to be applied to (23): If $\chi(p'')=0$, the transition density $\rho_{p'p''}(q)$ is still

$$(23') \qquad \rho_{p'p''}(q) = \chi^*(p')\, e^{\frac{2\pi i}{h}(p''-p')q}$$

for transitions from p' to an empty state p'' as if $\chi(p'')$ were unity. This explains the fact that in addition to the *coherent* diffraction of a double ray produced by transitions $+p \to -p$ there is an incoherent scattering of light in all directions, due to photons that are deflected by matter transitions $\pm p \to p'$, where p' represents a new momentum not originally present in the double ray itself.

§ 16. RESULTANT VALUES OF PHYSICAL FUNCTIONS; MATRIX ELEMENTS

Let us consider a "multiple ray" characterized by a certain abundance function $\sigma(p)$ for various momenta p. We learned in § 15 that deflections of photons due to transition processes $p' \to p''$ in the matter occur with a probability proportional to $\sigma(p').\sigma(p'')$. The same optical phenomenon could be explained as indicating a distribution of Huygens sources with the density and phase (23):

$$(26) \qquad \rho_{p''p'}(q) = \chi(p')\, e^{\frac{2\pi i}{h}p'q}\, \chi^*(p'')\, e^{-\frac{2\pi i}{h}p''q}.$$

Now consider a physical function $f(q)$ of the co-ordinate q. (For example, the electric moment $e.q$ of a charged mass point in the position q, or its moment of inertia mq^2, or any other physical quantity.) If $\rho(q)$ is the density at the point q, then $f(q)$ will have the resultant value for the whole matter ray

$$<f> = \int \rho(q) f(q)\, dq. \tag{27}$$

If we observe in particular secondary photons deflected in a direction belonging to a matter transition $p' \to p''$, then the corresponding wave density ρ will be the $\rho_{p'p''}$ of (23) and f will appear to have the "transition value"

$$\begin{aligned} <f>_{p'p''} &= \int \rho_{p'p''}(q) f(q)\, dq \\ &= \chi(p')\chi^*(p'') \int e^{\frac{2\pi i}{h}(p'-p'')q} f(q)\, dq. \end{aligned} \tag{28}$$

The two factors $\chi(p')$ and $\chi^*(p'')$ pertain to the special ray with special abundances $\sigma(p')$ and $\sigma(p'')$. The last factor

$$\int e^{\frac{2\pi i}{h}(p'-p'')q} f(q)\, dq = f_{p'p''} \tag{29}$$

is called the *matrix element* of $f(q)$ with respect to the transition $p' \to p''$. The matrix element $f_{p'p''}$ is independent of the accidental abundances $\sigma(p')$ and $\sigma(p'')$. It is the resultant of f pertaining to abundances $\sigma' = 1$ and $\sigma'' = 1$. Later on we have to generalize the definition of a matrix element to the case in which f is a function $f(q,p)$ of co-ordinates *and* momenta.

§ 17. PULSATING DENSITY

The rest of this part will be devoted to applying our former results to the case of *time* instead of *space*. Let a very small piece of matter be illuminated with light of the frequency ν_0. Suppose we observe that the light after being reflected by, or transmitted through, the matter has acquired two additional spectral components $\nu_0 + \nu_1$ and $\nu_0 - \nu_1$. We would infer that the matter density is periodic in time with the frequency ν_1. Indeed, if ρ at the illuminated point is pulsating in the manner

$$(30) \qquad \rho(t) = \bar{\rho} \, . \, 2\cos^2\left(2\pi\frac{\nu_1}{2}t\right) = \bar{\rho} \, . \, [1+\cos(2\pi\nu_1 t)],$$

then it will serve as a Huygens centre stimulated by the incident amplitude $\cos 2\pi\nu_0 t$ to produce the secondary amplitude

$$A(t) = \text{const.}\,\rho(t)\cos(2\pi\nu_0 t)$$
$$= \text{const.}\,\bar{\rho}\,\{\cos(2\pi\nu_0 t) + \tfrac{1}{2}\cos 2\pi(\nu_0+\nu_1)t + \tfrac{1}{2}\cos 2\pi(\nu_0-\nu_1)t\}.$$

The observed light will thus consist of the three spectral components ν_0 and $\nu_0 \pm \nu_1$. We can instead interpret these secondary spectral lines by means of photons incident with the energy $E_0 = h\nu_0$ and changing their energy to $E' = h(\nu_0+\nu_1)$ and $E'' = h(\nu_0-\nu_1)$ upon colliding with the matter. The conservation rule tells us then that the particles of matter must lose or gain energy amounts $\epsilon_1 = h\nu_1$ in making transitions between two energy levels ϵ' and ϵ'' that have the difference $\epsilon' - \epsilon'' = \epsilon_1$. The absolute values of ϵ' and ϵ'' have no physical significance.

By this twofold interpretation of the same optical phenomenon we find that the matter at the observed point must have two complementary properties. First, its density ρ must be pulsating in the manner (30). Second, there must be an abundance of two corpuscular energies

$$(31) \qquad \sigma(\epsilon') = \sigma(\epsilon''),$$

where $\epsilon' - \epsilon'' = \epsilon_1 = h\nu_1$. Equation (31) means that energy changes of only the amount $|\epsilon' - \epsilon''| = \epsilon_1$ are within the scope of the observed state of matter without transforming it into a new state.

§ 18. GENERAL RELATION BETWEEN $\rho(t)$ AND $\sigma(\epsilon)$

The direct formal relation between the density function $\rho(t)$ and the abundance function $\sigma(\epsilon)$ can be derived as follows. In the example of the periodic density of (30) we notice that $\rho(t)$ is the absolute square of the density amplitude

$$\psi(t) = \frac{\bar{\rho}}{\sqrt{2}}\left(e^{\frac{2\pi i}{h}\epsilon' t} + e^{\frac{2\pi i}{h}\epsilon'' t}\right), \quad \text{where } \epsilon' - \epsilon'' = \epsilon_1 = h\nu_1.$$

ψ represents a superposition of two vibrations of the same amplitude. In a more general case, several energies ϵ', ϵ'', ... may be present with the abundances

$$\sigma(\epsilon')=\sigma', \quad \sigma(\epsilon'')=\sigma'', \ldots$$

and with the abundance amplitudes

$$\chi(\epsilon')=\sqrt{\sigma'}e^{i\delta'}, \quad \chi(\epsilon'')=\sqrt{\sigma''}e^{i\delta''}.$$

$\psi(t)$ is then obtained as the Fourier series (cf. (15))

$$(32) \qquad \psi(t)=\Sigma\chi(\epsilon')e^{\frac{2\pi i}{h}\epsilon' t}=\Sigma\sqrt{\sigma'}e^{\frac{2\pi i}{h}\epsilon' t+i\delta'}.$$

The density $\rho=|\psi|^2$ becomes finally

(32′)

$$\rho(t)=\sigma'^2+\sigma''^2+\ldots+\Sigma\sqrt{\sigma'\sigma''}\cos\left[\frac{2\pi}{h}(\epsilon'-\epsilon'')t+(\delta'-\delta'')\right]+\ldots,$$

which consists of constant terms plus periodic "interference" terms.

If the abundance is a continuous function of ϵ, then we must represent $\psi(t)$ as the Fourier integral (cf. (16))

$$(33) \qquad \psi(t)=\int_{-\infty}^{\infty}\chi(\epsilon)e^{\frac{2\pi i}{h}\epsilon t}\,dt.$$

The inverse of this integration is

$$(33') \qquad \chi(\epsilon)=\frac{1}{h}\int_{-\infty}^{\infty}\psi(t)e^{-\frac{2\pi i}{h}\epsilon t}\,dt.$$

§ 19. TRANSITION DENSITY; MATRIX ELEMENTS

The same consideration which led to the "transition density" $\rho_{p''p'}(x)$ of (23) can be applied with slight alterations to the case of vibrations ν_0 of light or to photons $E_0=h\nu_0$ which are transformed into vibrations $\nu_0\pm\nu_1$ or photons of the energy $E=h(\nu_0\pm\nu_1)$. The light waves $\nu_0\pm\nu_1=\nu$ will appear to originate from a secondary Huygens source of the pulsating "transition density"

$$(34) \qquad \rho_{\epsilon''\epsilon'}(t)=\chi(\epsilon')e^{\frac{2\pi i}{h}\epsilon' t}\cdot\chi^*(\epsilon'')e^{-\frac{2\pi i}{h}\epsilon'' t} \quad \text{(cf. (23))},$$

where ϵ' and ϵ'' are two energy values of a particle of matter.

They satisfy the relation

$$\text{(34')} \qquad \frac{\epsilon' - \epsilon''}{h} = \nu_1 \quad \text{(cf. (21))}.$$

At the same time, if we use a corpuscular interpretation, it would appear that incident photons of the energy E_0 are transformed into photons of energy $E_0 \pm E_1$ due to transition processes $\epsilon' \rightarrow \epsilon''$ in the matter at a rate proportional to

$$\text{(35)} \qquad \sigma(\epsilon')\sigma(\epsilon'') \quad \text{(cf. (24))},$$

where ϵ', ϵ'', E_0 and E_1 are related by the energy conservation rule

$$\text{(35')} \qquad \epsilon' - \epsilon'' = E_1.$$

By analogy with (27), the resultant of any physical function $f(t)$ will appear to have the value

$$<f> = \int \rho(t) f(t)\, dt$$

when observed optically and interpreted according to the wave theory of light. In particular, those diffracted light waves which correspond to deflected photons according to the energy relation (34') seem to indicate a resultant value of f given by

$$\text{(36)} \qquad <f>_{\epsilon''\epsilon'} = \chi(\epsilon')\chi^*(\epsilon'').f_{\epsilon''\epsilon'} \quad \text{(cf. (28))},$$

where

$$\text{(37)} \qquad f_{\epsilon''\epsilon'} = \int e^{\frac{2\pi i}{h}\epsilon' t} . e^{-\frac{2\pi i}{h}\epsilon'' t} f(t)\, dt \quad \text{(cf. (29))}.$$

This is called the "matrix element" of the physical function $f(t)$ with respect to the transition $\epsilon' \rightarrow \epsilon''$. It is the resultant value of f pertaining to the abundances $\sigma(\epsilon') = 1 = \sigma(\epsilon'')$.

The matrix elements are typical for quantum mechanics in so far as they express a relation between two interpretations of the same state. According to the corpuscular theory matter can be in a transitional state $\epsilon' \rightarrow \epsilon''$. According to the theory of waves the same state displays a certain density $\rho_{\epsilon'\epsilon''}$ and a certain value $f_{\epsilon'\epsilon''}$ of the physical quantity f. In Parts III and IV we shall arrive at a more general method of calculating the resultant values of physical functions $f(q, p)$ in more general "states of transition" for particles in force fields or waves in inhomogeneous media.

PART II

THE PRINCIPLE OF UNCERTAINTY

§ 20. OPTICAL OBSERVATION OF DENSITY IN MATTER PACKETS

Heisenberg's (10) principle of uncertainty can be developed as a special application of the general theory of observation of Part I.

Let a train of parallel monochromatic light be incident at an angle α_0 upon a certain unknown distribution of matter along the q-axis. Suppose we observe that the light is diffracted over a range $\Delta\alpha$ of angles around the original direction α_0 in the form of an intensity maximum, say in the form of a Gaussian error curve,† so that the amplitude observed in the direction of α is

$$A(\alpha) = \text{const.}\, e^{-\frac{1}{4}\left(\frac{\alpha-\alpha_0}{\Delta\alpha}\right)^2},$$

with $\Delta\alpha$ as "half-width". It is more convenient to plot the amplitude as a function of $\cos\alpha$. If $\Delta\alpha$ is small, we may write

$$\frac{\alpha-\alpha_0}{\Delta\alpha} = \frac{\cos\alpha - \cos\alpha_0}{\Delta(\cos\alpha)}.$$

The observed amplitude may thus be written in the form

$$A(\alpha) = \text{const.}\, e^{-\frac{1}{4}\left(\frac{\cos\alpha-\cos\alpha_0}{\Delta(\cos\alpha)}\right)^2}, \tag{1}$$

where $\Delta(\cos\alpha)$ is the half-width in the $\cos\alpha$ diagram (Fig. 6).

Our task is then to make inferences regarding the source of this diffracted light. Using the wave theory of light, we can determine the density distribution of the matter along the q-axis. Using the photon theory, we can determine the momenta given out by the matter. In order to obtain quantitative results we proceed as follows:

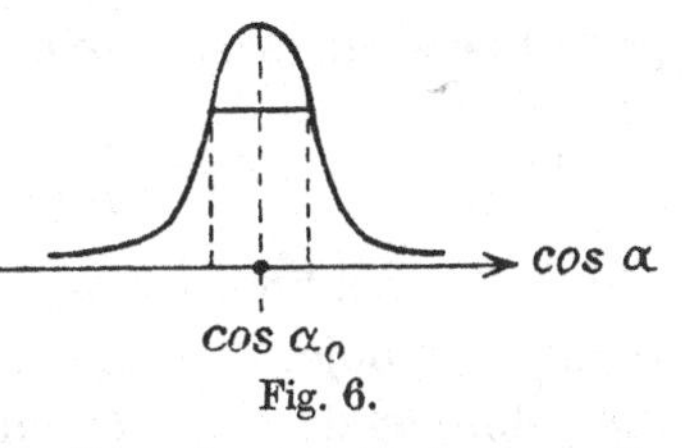

Fig. 6.

† See footnote on p. 41.

Let us assume that the matter which is the source of that diffraction is crowded together about a point q_0 in the form of a maximum of the width Δq, such that the density $\rho(q)$ is represented by

$$\rho(q) = \text{const.}\, e^{-\left(\frac{q-q_0}{\Delta q}\right)^2}. \tag{2}$$

Observing the width $\Delta\alpha$ or $\Delta(\cos\alpha)$ we are able, with the help of the *wave theory* of light, to calculate the width Δq of the matter causing the diffraction. We may apply the general formula of the Huygens principle:

$$A(\alpha) = \text{const.} \int \rho(q)\, e^{i.\phi_\alpha(q)}\, dq, \tag{3}$$

where the phase $\phi_\alpha(q)$ of the light diffracted at the point q into the direction of α is†

$$\phi_\alpha(q) = \frac{2\pi}{\Lambda} q(\cos\alpha - \cos\alpha_0),$$

and where $\rho(q)$ is given by (2). So we obtain the diffracted amplitude

$$A(\alpha) = \text{const.} \int_{-\infty}^{\infty} e^{-\left(\frac{q-q_0}{\Delta q}\right)^2 + \frac{2\pi i}{\Lambda} q(\cos\alpha - \cos\alpha_0)}\, dq. \tag{4}$$

The integration can be carried out mathematically (in the same manner as shown later in (13′)), and gives the result

$$A(\alpha) = \text{const.}\, e^{-\frac{1}{4}\left[\frac{2\pi}{\Lambda}(\cos\alpha - \cos\alpha_0)\Delta q\right]^2}. \tag{5}$$

Comparing (5) with (1) we find: The angular width $\Delta(\cos\alpha)$ of the diffraction maximum and the width Δq of the diffracting matter packet bear the relation to each other

$$\Delta q = \frac{\Lambda}{2\pi} \frac{1}{\Delta(\cos\alpha)} = \frac{\Lambda}{2\pi} \frac{1}{\sin\alpha_0 \Delta\alpha}. \tag{6}$$

This is the well-known equation for the *optical resolving power*: The smaller the object Δq, the larger is the angular spread $\Delta\alpha$ of the light diffracted from it (regardless of whether there is really no matter outside of Δq or whether only the matter inside of Δq is

† Implicitly we assume at this point that the matter does not contribute any phase jumps to the incident light.

illuminated by primary light waves through a shutter with a hole Δq).

§ 21. DISTRIBUTION OF MOMENTA IN MATTER PACKETS

Let us now explain the same optical diffraction through the angle $\Delta\alpha$ around α_0 by means of *photons* of momentum $P = h/\Lambda$. The q-component of the momentum, $P . \cos\alpha_0$, appears to be changed into $P . \cos\alpha$ owing to collisions with the matter in which the latter gives off corresponding impulses

(7) $p'' - p' = P(\cos\alpha - \cos\alpha_0)$ (conservation of momentum)

by means of hypothetical "transition processes" $p' \to p''$. In § 15 of Part I we learned that the probability amplitude of such a transition process (responsible for the amplitude of the deflected light) is proportional to the product of the abundance amplitudes $\chi(p')$ and $\chi(p'')$ according to the formula

$$A(\alpha) = \text{const.}\, \chi(p')\, \chi^*(p'').$$

There may, however, be many pairs p' and p'' satisfying (7) which cause a photon to be deflected into the same direction α. So we must write

(8) $$A(\alpha) = \text{const.} \int \chi(p')\, \chi^*(p'')\, dp',$$

where $$p'' = p' + P(\cos\alpha - \cos\alpha_0).$$

We try now to find the form of the abundance amplitude $\chi(p)$ which renders (8) identical with the observed maximum (1). Since $A(\alpha)$ is condensed within a small range $\Delta\alpha$ around α_0 mainly, we infer that only small differences $p' - p''$ are responsible for these small deflections. So we suspect only a small range of contributing momenta in the form of the trial function

(9) $$\chi(p) = \text{const.}\, e^{-\frac{1}{2}\left(\frac{p-p_0}{\Delta p}\right)^2},$$

(9′) $$\sigma(p) = \text{const.}\, e^{-\left(\frac{p-p_0}{\Delta p}\right)^2},$$

where Δp represents the half-width of the range of momenta. [We have failed here to add a phase factor in $\chi(p)$. This is equivalent to perfectly random distribution or lack of preference of any such phase.] Our task is then only to determine the magnitude of the range Δp over which the momenta of the matter are distributed in order to give the observed result (1). So we must insert (9) into (8) and compare the result with (1). We obtain first from (8) and (9)

$$A(\alpha) = \text{const.} \int_{-\infty}^{\infty} e^{-\frac{1}{2}\left(\frac{p'-p_0}{\Delta p}\right)^2} e^{-\frac{1}{2}\left(\frac{p'-p_0+P(\cos\alpha-\cos\alpha_0)}{\Delta p}\right)^2} dp'.$$

In order to evaluate this integral we write the sum of the two exponents in the form

$$-\frac{1}{4}\left(\frac{P(\cos\alpha-\cos\alpha_0)}{\Delta p}\right)^2 - \left(\frac{p'-p_0}{\Delta p} + \frac{P(\cos\alpha-\cos\alpha_0)}{2\Delta p}\right)^2.$$

Calling the second bracket u, we have $du = (1/\Delta p)\,dp'$ and

$$A(\alpha) = \text{const.}\, e^{-\frac{1}{4}\left(\frac{P(\cos\alpha-\cos\alpha_0)}{\Delta p}\right)^2} \int_{-\infty}^{\infty} e^{-u^2} du.$$

Since the last factor, the integral, is constant and independent of α, we obtain

$$A(\alpha) = \text{const.}\, e^{-\frac{1}{4}\left(\frac{P(\cos\alpha-\cos\alpha_0)}{\Delta p}\right)} \tag{10}$$

for the diffracted amplitude produced by photons deflected by matter of abundance (9′). Comparing (10) with the observed result (1), we obtain the following relation between the angular spread $\Delta\alpha$ and the half-width Δp of the abundance distribution

$$\Delta p = P.\Delta(\cos\alpha) = P\sin\alpha_0 . \Delta\alpha. \tag{10′}$$

According to § 20 the same angle $\Delta\alpha$ indicated, as inferred from the wave theory of light, that the matter is crowded together with the half-width (6)

$$\Delta q = \frac{\Lambda}{2\pi}\frac{1}{\Delta(\cos\alpha)}.$$

Identifying P with h/Λ we obtain from (6) and (10′) the following relation between Δp and Δq:

$$\Delta p . \Delta q = \frac{h}{2\pi}. \tag{11}$$

This formula connects the density function (2) with the complementary abundance function (9′). The same optical diffraction which indicates, according to the wave theory, that matter is spread over a range Δq, tells us according to the corpuscular theory of light that the momenta of the matter are distributed over a range Δp. The latter statement means that the matter is capable of giving out impulses $|p'-p''|$ that are not of larger order than Δp. The absolute values of the momenta p' and p'' cannot be inferred by observing the deflection of photons, and it is only their difference that has a physical meaning in this connection.

§ 22. MATHEMATICAL RELATION BETWEEN ρ AND σ

Before we discuss the physical significance of the "uncertainty relation" (11) we may now derive this relation by means of the direct mathematical method developed in § 12 of Part I, without reference to optical observations.

Suppose the density distribution of matter along the q-axis to have the form of a Gaussian error curve†

$$\rho(q) = e^{-\left(\frac{q}{\Delta q}\right)^2}, \tag{12}$$

putting its peak at $q=0$ for the sake of simplicity. From ρ we obtain the density amplitude

$$\psi(q) = e^{-\frac{1}{2}\left(\frac{q}{\Delta q}\right)^2}, \tag{12'}$$

in which we drop the arbitrary phase factor $e^{i\beta(q)}$ altogether, assuming a completely random distribution of the phase $\beta(q)$. In order to obtain the abundance amplitude $\chi(p)$ we use the general formula (16) of Part I:

$$\chi(p) = \frac{1}{h}\int_{-\infty}^{\infty} \psi(q)\, e^{-\frac{2\pi i}{h}p.q}\, dq. \tag{13}$$

In our present case (12′), we have

$$\chi(p) = \frac{1}{h}\int_{-\infty}^{\infty} e^{-\frac{1}{2}\left(\frac{q}{\Delta q}\right)^2 - \frac{2\pi i}{h}p.q}\, dq. \tag{13'}$$

† Heisenberg (11) has proved by a variational method that this Gaussian form is the most favourable one for restricting the range of Δp. All other forms of the density maximum give $\Delta p > \frac{h}{2\pi}\frac{1}{\Delta q}$.

In order to evaluate this integral we write the exponent in the form

$$-\frac{1}{2}\left(\frac{2\pi}{h}p\Delta q\right)^2-\frac{1}{2}\left(\frac{q}{\Delta q}+\frac{2\pi i}{h}p\Delta q\right)^2.$$

Calling the second bracket u, we have $du=(1/\Delta q)\,dq$ and

$$\chi(p)=\frac{\Delta q}{h}.e^{-\frac{1}{2}\left(\frac{2\pi i}{h}\right)p\Delta q^2}\int e^{-\frac{1}{2}u^2}du.$$

Since the last integral† has the constant value $\sqrt{2\pi}$, we obtain

$$\chi(p)=\sqrt{2\pi}\frac{\Delta q}{h}.e^{-\frac{1}{2}\left(\frac{2\pi}{h}p\Delta q\right)^2} \tag{14}$$

and finally

$$\sigma(p)=2\pi\left(\frac{\Delta q}{h}\right)^2.e^{-\left(\frac{2\pi}{h}p\Delta q\right)^2}. \tag{15}$$

(15) represents a Gaussian error curve of the form

$$\sigma(p)=\text{const.}\,e^{-\left(\frac{p}{\Delta p}\right)^2}, \tag{16}$$

if Δp is written for

$$\Delta p=\frac{h}{2\pi}\frac{1}{\Delta q}, \tag{17}$$

which is again the relation (11).

We may generalize this result for a three-dimensional matter packet, whose density is described by

$$\rho(xyz)=\text{const.}\,e^{-\left(\frac{x-x_0}{\Delta x}\right)^2-\left(\frac{y-y_0}{\Delta y}\right)^2-\left(\frac{z-z_0}{\Delta z}\right)^2}. \tag{18}$$

The complementary abundance function of momenta then becomes

$$\sigma(p_xp_yp_z)=\text{const.}\,e^{-\left(\frac{p_x-p_x^0}{\Delta p_x}\right)^2-\left(\frac{p_y-p_y^0}{\Delta p_y}\right)^2-\left(\frac{p_z-p_z^0}{\Delta p_z}\right)^2}, \tag{18'}$$

where the relation between the half-widths is now

$$\Delta p_x.\Delta x=\frac{h}{2\pi},\quad \Delta p_y.\Delta y=\frac{h}{2\pi},\quad \Delta p_z.\Delta z=\frac{h}{2\pi}. \tag{19}$$

† The limits of the integral are now $u=\pm\frac{\infty}{\Delta q}+\frac{2\pi i}{h}p.\Delta q$, but their absolute values are still $\pm\infty$.

We note that the inverse of (13),

$$\psi(q)=\int \chi(p) \, . \, e^{\frac{2\pi i}{h} p.q} \, dp,$$

represents a superposition of various "wave functions"

$$\chi\left(\frac{h}{\lambda}\right) e^{\frac{2\pi i}{\lambda} q} \quad \text{with} \quad \lambda=\frac{h}{p},$$

whose amplitudes are of appreciable magnitude only for wave lengths $\lambda=h/p$ within an interval

$$\Delta\left(\frac{1}{\lambda}\right)=\frac{1}{h}\Delta p=\frac{1}{2\pi}\frac{1}{\Delta q}.$$

$\rho(q)$ represents a *wave packet*, and $\sigma(p)$ a packet of momenta.

The definition of the width of a Gaussian error curve is rather arbitrary. For instance, we may introduce new half-widths

$$\text{(20)} \qquad \delta x=\frac{\Delta x}{\sqrt{2\pi}}, \quad \delta p_x=\frac{\Delta p_x}{\sqrt{2\pi}}, \ldots,$$

which then satisfy the relations

$$\text{(20')} \qquad \delta p_x \, . \, \delta x=h, \quad \delta p_y \, . \, \delta y=h, \quad \delta p_z \, . \, \delta z=h.$$

§ 23. CAUSALITY

The relation
$$\Delta p \, . \, \Delta q=\frac{h}{2\pi}$$

connects two hypothetical properties of the same piece of matter: its density distribution over a width Δq and its ability to transfer momenta up to an amount Δp, both properties judged from optical observations in their dual interpretation. These properties are complementary [in the same sense as the two properties of an infinite crystal, viz. (1) it is a system with a periodicity λ of its density in space; (2) it is a system that gives out impulses which are multiples of certain fundamental momenta $p=h/\lambda$].

Let us begin the discussion of the relation $\Delta p_x \, . \, \Delta x=h/2\pi$ with a physical example. Consider a ray of matter falling on a screen with a hole of the width Δx. Just behind the hole, then, the matter is confined wholly within the range Δx. For the sake of mathematical simplicity we may assume that the edges of the hole are not absolutely opaque, so that the matter density of our ray just behind the hole is given by

$$\rho(x)=\rho_0 e^{-\left(\frac{x}{\Delta x}\right)^2},$$

with the half-width Δx and the peak at $x=0$. Although the ray has no transverse component of momentum in front of the screen, it presents behind it a range Δp_x of momenta and a corresponding range of transverse velocities

$$\Delta v_x=\frac{1}{m}\Delta p_x=\frac{h}{2\pi m}\frac{1}{\Delta x}$$

according to the uncertainty principle (19).

So the matter ray will be spread out from the original direction $\alpha_0=90°$ into a bundle of directions

$$\Delta\alpha=\frac{\Delta v_x}{v}=\frac{\Delta p_x}{p}=\frac{h}{2\pi p}\cdot\frac{1}{\Delta x},$$

if p is the total momentum of the incident matter particles.

It has often been said that this acquisition by the matter particles (or photons) of a transverse component when going through a *hole* constitutes a break with the rules of classical mechanics, in particular with the fundamental principle of *cause* and *effect*: A hole, that is, the absence of screen material, should have no effect whatever. Instead of travelling straight ahead within the boundaries of the geometrical shadow, the particles seem to possess the power of knowing about the rules of wave diffraction, to which they submit not individually, but in the average, so as to produce the intensity pattern predicted by the wave theory. It is therefore not surprising that philosophers should have become interested in this apparent contradiction of the fundamental principles.

But there is no reason for being alarmed when matter or light is diffracted into a bundle of directions by a hole, or produces interference fringes when passed through a regular arrangement of holes (grating). If the picture of particles is adhered to, then there *is* a mechanical causal explanation: The screen with a hole *has* a hole only if viewed and interpreted from the wave-theoretical point of view. From the corpuscular standpoint "the screen with a hole" represents an arrangement of particles that is capable of transferring momenta of the order of $\Delta p_x = \frac{h}{2\pi}\frac{1}{\Delta x}$ strictly in accordance with the conservation rule. (Its capability of diffraction is, by the way, exactly the same as that of a peg fitting into the hole, a result known in optics as Babinet's theorem.) And a screen which from the standpoint of waves appears to contain a number of holes (or pegs) at equal distances a will appear from the standpoint of particles to be a system capable of transferring momenta $p = n.h/a$ to incident photons with conservation of the total momentum, again in a perfectly causal way.†

Similar considerations may be applied in order to explain the diffused maxima that are produced by a finite grating of N lines or by an infinite grating of which only N lines are illuminated. The wave theory explains immediately that only the N *illuminated* grating lines participate in the production of the interference maxima. In order to explain the same facts from the corpuscular point of view we must say that only the *illuminated* part of the grating reacts to the incident photons and transfers momentum to them. The probability of various impulses $p' - p''$ being transmitted is proportional to $|\chi(p') \cdot \chi^*(p'')|^2$ according to § 15. But in calculating $\chi(p)$ according to (13) we must now use a density amplitude $\psi(q)$ which represents the density along only N *illuminated* grating lines, $\psi(q)$ being zero outside of them. Hence $\chi(p)$ now becomes quite different from the abundance amplitude of an infinite grating. In this way Epstein and Ehrenfest (12) were

† We do not mean that from the corpuscular standpoint a system of artificial holes or grooves on a plate are nothing but a manufactured spectrum of momenta $\sigma(p)$. Our former considerations refer to free particles. In the plate there are mutual forces. The grooves and holes contribute additional impressed forces (cf. end of § 10). Furthermore, a given *macroscopic* formation of matter can always be described *completely* in wave terms as well as in corpuscular terms, the uncertainty being on a much smaller scale (cf. footnote, p. 24).

able to explain the details of the diffraction pattern of a finite grating by means of the corpuscular picture. The result was the same as that of the wave theory. In fact the whole mathematical procedure of the corpuscular theory is quite complementary to that of the wave theory, and only the words and the interpretations used are different. Compare for instance the procedure of § 20 with that of § 21.

§ 24. UNCERTAINTY

Returning to the example of a screen with a hole Δx, we learned that the same screen in corpuscular terms is a system that gives out impulses of the order $\Delta p_x = \frac{h}{2\pi}\frac{1}{\Delta x}$. This was used to explain in a corpuscular causal way the diffraction of incident photons P into a bundle of directions $\Delta\alpha = \frac{h}{2\pi P}\frac{1}{\Delta x}\frac{1}{\sin\alpha_0}$. The diffraction was thus attributed to the *instrument* which confined the waves to the width Δx or gave out pushes Δp_x. The same fact may be expressed by saying that the ray, by virtue of its being confined to the width Δx, possesses the *inherent* quality of momenta spread over the interval Δp_x. Or again: An individual particle belonging to that ray possesses a position *uncertain* within amount Δx, and a momentum *uncertain* within amount Δp_x.

Take as another example a large grating with the distance a between successive lines. A parallel train of light (or matter) after having passed through the grating is said to have the "inherent quality" of containing momenta $P_x = P_x^0 + n.h/a$, either by virtue of its own periodic structure, or because momentum $n.h/a$ is acquired from the grating. In terms of corpuscles it remains then *uncertain* whether an individual particle will acquire the additional momentum $1.h/a$, or $2.h/a$, etc. The uncertainty consists here in the number n, that is the "order" of diffraction into which an individual particle will be sent by the grating.

§ 25. UNCERTAINTY DUE TO OPTICAL OBSERVATION

Heisenberg has illustrated the uncertainty relation

$$\Delta p_x \Delta x = \frac{h}{2\pi}$$

from the corpuscular point of view by an example in which the location x of an object is to be determined within an accuracy Δx. If we wish to determine the position x with the accuracy Δx, we can do this by scanning the x-axis with a microscope whose objective has the width Δx. Now if we use light of the wave length Λ, then the natural diffraction through the objective Δx requires that we observe with an eyepiece whose angular aperture viewed from the objective is at least

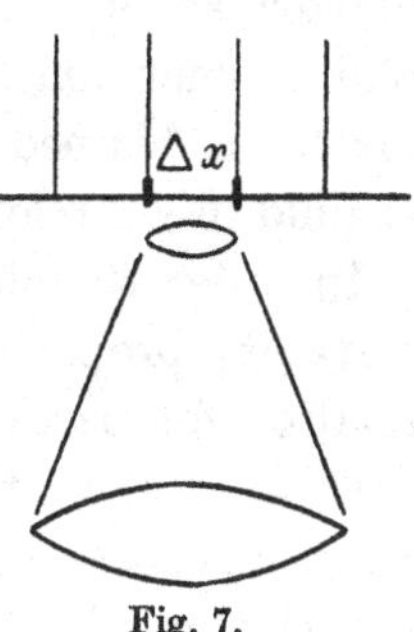

Fig. 7.

$$\Delta\alpha = \frac{\Lambda}{2\pi}\frac{1}{\Delta x},$$

according to the rule (6) of the optical resolving power. Hence photons P come into our eye that have acquired x-momentum up to the amount

$$P \,.\, \cos(90° + \Delta\alpha) - P = P \,.\, \Delta\alpha.$$

The recoil Δp_x which the object itself may have obtained from a deflected photon is just as large $\Delta p_x = P \,.\, \Delta\alpha$. Owing to the very act of observing the object within the limits of Δx, its momentum becomes uncertain within limits Δp_x. Comparing the last two equations, we have then

$$\Delta p_x = P \,.\, \Delta\alpha = P\frac{\Lambda}{2\pi}\frac{1}{\Delta x},$$

and since $P \,.\, \Lambda = h$, we obtain finally

$$\Delta x \Delta p_x = \frac{h}{2\pi}.$$

We cannot determine then by an optical observation the position x and the momentum p_x of a particle more accurately than within the two reciprocal uncertainties Δx and Δp_x whose product is $h/2\pi$.

This however is not a very satisfactory deduction of the uncertainty relation. It tells us only that one cannot determine accurately both position and momentum of a particle of matter, since one cannot determine with certainty the place and momentum of a deflected photon. The uncertainty of matter is blamed on the uncertainty of light, and one moves around in a circle.

In order therefore to avoid any direct reference to the uncertainty properties of *light*, we may deduce the uncertainty relation for particles of matter in the following manner: If we wish to measure the momentum p_x of a particle of matter we may superpose on it a homogeneous matter ray with known momentum p_x^0, to which belongs a *constant* density distribution $\rho_0(x) = \text{const}$. The presence of our matter particle p_x will then produce a "beat" of intensity, the number of intensity maxima per unit of length being

$$n = \frac{1}{\lambda} - \frac{1}{\lambda_0} = \frac{p_x - p_x^0}{h}.$$

In order to distinguish this number from a possibly different beat number

$$n + \Delta n = \frac{(p_x + \Delta p_x) - p_x^0}{h},$$

which would belong to a matter particle of the momentum $p_x + \Delta p_x$, we would have to observe a section of length

$$\Delta x = \frac{1}{\Delta n} = \frac{h}{\Delta p_x}$$

along the x-axis. Thus the place where an individual particle displays the momentum p_x with the uncertainty Δp_x will remain uncertain by $\Delta x = h/\Delta p_x$.

The uncertainty principle gives a definite range of accuracy beyond which the concepts of location and momentum can no longer be applied. Two pairs of values (p', x') and (p'', x'') that fall within a rectangular range of area $h/2\pi$ in an x-p_x-diagram (Fig. 8) cannot be distinguished from each other. Only so long as we allow for uncertainties as large as Δx and Δp_x with the product $h/2\pi$ are we entitled to describe physical phenomena in terms of corpuscles. It is due only to the smallness of the quantum h that the corpuscular picture has proved so successful in accounting for a great number of macroscopic physical facts.

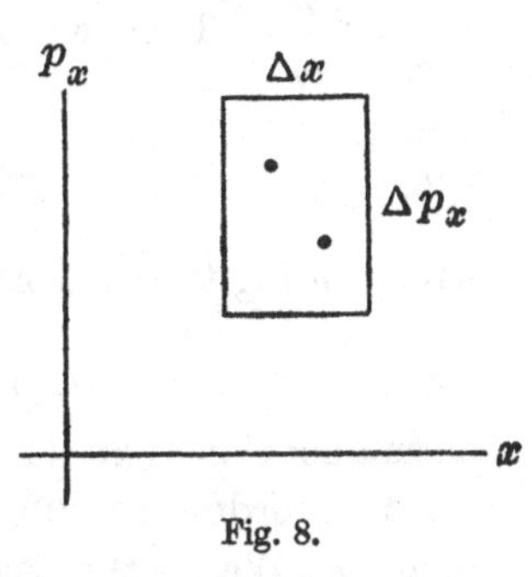

Fig. 8.

§ 26. DISSIPATION OF MATTER PACKETS; RAYS IN WILSON CHAMBER

Suppose a ray of matter travelling in the y-direction with the velocity v_y and with the momentum $p_y = mv_y$ to be confined by a shutter to the cross-section Δx_0. For reasons of simplicity suppose again that Δx_0 represents the "half-width" of the density just behind the shutter, the density at $y = 0$ being

$$\rho(x, 0) = \text{const.}\, e^{-\left(\frac{x}{\Delta x_0}\right)^2}. \tag{21}$$

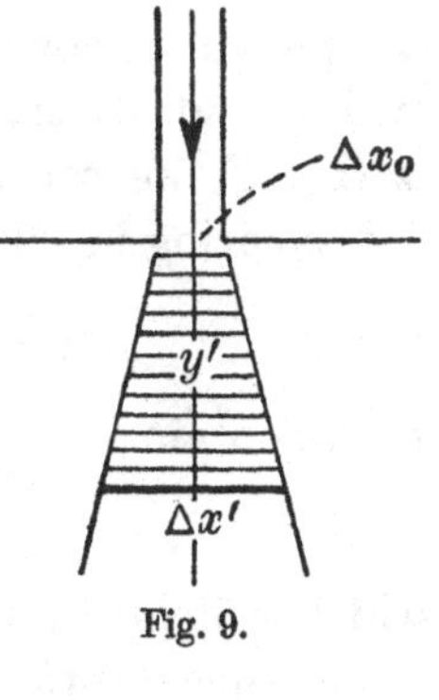

Fig. 9.

What is the density distribution at a distance y' below the shutter? Remembering that the ray will be spread over an angular aperture of half-width

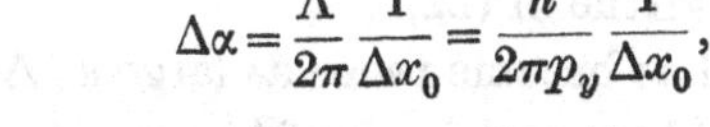

$$\Delta\alpha = \frac{\Lambda}{2\pi}\frac{1}{\Delta x_0} = \frac{h}{2\pi p_y}\frac{1}{\Delta x_0},$$

its half-width in the x-direction will be approximately

$$\Delta x' = y'\Delta\alpha = \frac{y'h}{2\pi p_y}\frac{1}{\Delta x_0}.$$

Instead of the distance y' let us introduce the travelling time $t' = \frac{y'}{v_y} = \frac{y'm}{p_y}$. Thus we obtain as the half-width $\Delta x'$, after a time t',

$$\Delta x' = \frac{ht'}{2\pi m}\frac{1}{\Delta x_0}, \tag{22}$$

corresponding to the density distribution

$$\rho(x, t') = \text{const.}\, e^{-\left(\frac{x}{\Delta x'}\right)^2}. \tag{22'}$$

The smaller the original half-width Δx_0, the faster will the matter spread, according to (22), giving rise to an ever-increasing half-width $\Delta x'$. We see that $\Delta x'$ after the time t' is quite independent of the y-component of the velocity; hence the spread of the matter originally confined to a half-width Δx will be the same even if $v_y = 0$.

We can explain this dissipation of matter packets also in the following manner: If $\rho(x, 0)$ of (21) is the density distribution, then there belongs to it a complementary abundance function of momentum:

$$\sigma(p_x) = \text{const.}\, e^{-\left(\frac{p_x}{\Delta p_x}\right)^2}, \tag{22''}$$

with a half-width $\Delta p_x = \frac{h}{2\pi}\frac{1}{\Delta x_0}$. This means, however, that there are present various velocities $v_x = (1/m)\,p_x$, so the matter packet will spread. If the original half-width Δx_0 is small compared with $\Delta x'$, we can obtain the density distribution at t' simply by replacing in (22'')

$$p_x \text{ by } mv_x = m\frac{x}{t'} \text{ and } \Delta p_x \text{ by } \frac{h}{2\pi}\frac{1}{\Delta x_0}.$$

So we obtain

$$\rho(x, t') = \text{const.}\, e^{-\left(\frac{mx}{t'}\cdot\frac{2\pi\Delta x_0}{h}\right)^2},$$

which is identical with (22') by virtue of (22).

The time τ within which $\Delta x'$ has become twice as large as Δx_0 is found from (22) by replacing t' by τ and $\Delta x'$ by $2\Delta x_0$, resulting in

$$\tau = \frac{4\pi m\,(\Delta x_0)^2}{h}.$$

On account of the smallness of h, this time is comparatively long, unless Δx_0 is very small. Take, for example, an α-ray with $m = 6{\cdot}6 \times 10^{-24}$ gr. that is sent through a shutter of the diameter $\Delta x_0 = 10^{-5}$ cm. (limit of optical visibility). Then τ becomes about 10^{-6} sec. If the α-ray travels with a velocity of $c/10 = 3 \times 10^9$ cm./sec., it covers during this time a distance of 30 metres. So the wave theory does not object to α-rays in a Wilson Chamber displaying straight linear paths (as long as they are not deflected by external forces) in spite of the diffraction through the hole. It will be different for electrons, whose mass m is almost 8000 times smaller than that of an α-particle. On the other hand, the shutters Δx used in practice are necessarily much greater than 10^{-5} cm., so that deviations from straight linear paths for electrons will not easily be observed either, unless they are subject to external forces.

§ 27. DENSITY MAXIMUM IN TIME

The foregoing results concerning space co-ordinates and momenta can immediately be transformed into results for time t and energy ϵ. Suppose the density of matter at a fixed point of space to be described by the Gauss function

$$\rho(t) = be^{-\left(\frac{t}{\Delta t}\right)^2}, \tag{23}$$

with a maximum at $t = 0$. Consider, for example, a shutter that is gradually opened to a matter ray and then closed again, the density ρ of the matter being observed at a fixed point behind the shutter. Or consider a constant current of matter that is illuminated during the time Δt by a light flash. The pulsation (23) corresponds to a certain abundance σ of various energies ϵ. In order to obtain $\sigma(\epsilon)$ we must first derive from (23) the density amplitude, containing the arbitrary phase $\beta(t)$:

$$\psi(t) = \sqrt{b}e^{i\beta(t)} . e^{-\frac{1}{2}\left(\frac{t}{\Delta t}\right)^2}, \tag{23'}$$

and then express the abundance amplitude as the Fourier integral

$$\chi(\epsilon)=\frac{1}{h}\int_{-\infty}^{\infty}\psi(t)\,e^{-\frac{2\pi i}{h}\epsilon t}\,dt.$$

Assuming a completely random distribution of the phase function $\beta(t)$, we obtain by analogy with (14) the result

$$\chi(\epsilon)=\sqrt{2\pi}\,b\,\frac{\Delta t}{h}\,.\,e^{-\frac{1}{2}\left(\frac{2\pi}{h}\epsilon\Delta t\right)^2},$$

and finally the abundance

$$\sigma(\epsilon)=\text{const.}\,e^{-\left(\frac{2\pi}{h}\epsilon\Delta t\right)^2}.$$

Introducing an interval $\Delta\epsilon$ with the help of the equation

$$\Delta\epsilon\,.\,\Delta t=\frac{h}{2\pi}, \tag{24}$$

we can write $\sigma(\epsilon)$ in the form

$$\sigma(\epsilon)=\text{const.}\,e^{-\left(\frac{\epsilon}{\Delta\epsilon}\right)^2}. \tag{25}$$

Thus $\Delta\epsilon$ proves to mean the half-width of the energy range. Instead of $\Delta\epsilon$ and Δt we may introduce $\delta\epsilon=\Delta\epsilon/\sqrt{2\pi}$ and $\delta t=\Delta t/\sqrt{2\pi}$. They satisfy the relation

$$\delta\epsilon\,.\,\delta t=h. \tag{25'}$$

§ 28. UNCERTAINTY OF ENERGY AND TIME

If we express the relation (24) $\Delta\epsilon\,.\,\Delta t=h/2\pi$ in corpuscular language, we obtain another example of Heisenberg's principle of uncertainty: The more one decreases the time interval Δt of a piece of matter being observed, the more does the range $\Delta\epsilon$ of the energies that are present in that piece of matter increase. Or in terms of a single particle: The smaller the uncertainty Δt of the time t at which an individual particle is observed, the larger the uncertainty $\Delta\epsilon$ of the energy value to be assigned to the particle.

These reciprocal uncertainties are usually explained as follows. Suppose we illuminate a point in space with monochromatic light ν_0. If we observe that at a certain time the light acquires an inhomogeneity of colour, we ascribe this to the presence of matter. The new colour means in corpuscular terms that the matter during its reaction with a photon must have changed the energy $h\nu_0$ of the photon to $h(\nu_0+\delta\nu)$. Thus the particle of matter must have changed in its own energy by $\epsilon'-\epsilon''=h.\delta\nu$, just at the time when its energy was to be measured. Hence the energy of the particle has become uncertain to the amount $h.\delta\nu$. If $\delta\nu$ is small, then the exact time at which that change of colour happened becomes all the more uncertain. Indeed, to tell a colour $\nu_0+\delta\nu$ from ν_0 we must observe at least during the time required for one *beat* between the two frequencies ν_0 and $\nu_0+\delta\nu$, that is, the time $\delta t \sim 1/\delta\nu$. This interval then represents the uncertainty of the time at which the matter particle is observed, and we have the relation

$$\delta t \sim \frac{1}{\delta\nu} = \frac{h}{\delta t} \quad \text{or} \quad \delta t.\delta\epsilon \sim h. \tag{26}$$

This consideration shifts the responsibility for the uncertainty in regard to particles of matter to uncertainties in regard to photons, leading round in a circle (see § 25). Instead we may argue as follows: In order to measure the energy ϵ of a particle of matter at a certain place we may superpose on it at that place other matter with the known characteristic energy ϵ_0 and hence with a constant density function $\rho_0(t)=\text{const}$. The presence of particles of matter ϵ will then be realized as a beat of the matter density with the beat frequency $\nu_1=(\epsilon-\epsilon_0)/h$. To tell this beat frequency from a possibly different beat frequency $\nu_1+\delta\nu=\dfrac{(\epsilon+\delta\epsilon)-\epsilon_0}{h}$ that would be produced if the particles of matter had the energy $\epsilon+\delta\epsilon$, a time interval $\delta t \sim \dfrac{1}{\delta\nu}=\dfrac{h}{\delta\epsilon}$ is required.

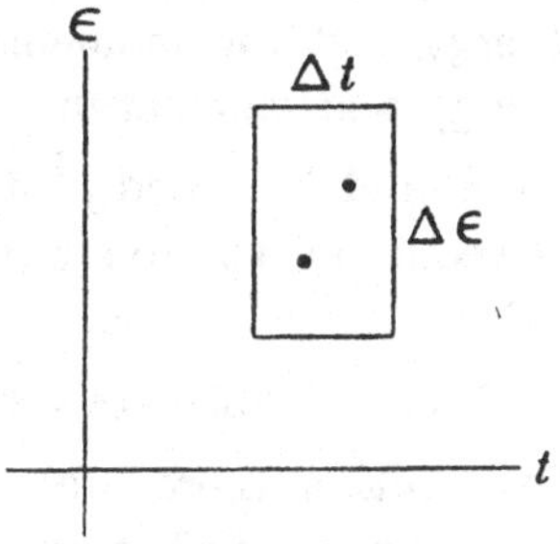

Fig. 10.

The uncertainty relation (24) gives a definite range beyond which it becomes meaningless to apply the concepts of corpuscles with certain energies at certain times. Two pairs of values ϵ', t' and ϵ'', t'' that fall within a rectangular section of the area $h/2\pi$ are indistinguishable (Fig. 10). Only so long as we allow uncertainties as large as $\Delta\epsilon$ and Δt, with the product $h/2\pi$, are we able and entitled to describe material phenomena in terms of corpuscles.

§ 29. COMPTON EFFECT; COMPTON-SIMON EXPERIMENT

The Compton effect deals with the change of colour of an incident X-ray if the latter is scattered by free electrons. First there is a *classical explanation* of the effect, worked out by O. Halpern (13): The incident light waves, by their radiation pressure, exert an accelerating force on the electrons and at the same time induce them to vibrate transversely. The electrons then emit a secondary radiation whose colour depends on the angle of observation, because of the Doppler shift. Compton and Debye gave a *corpuscular* explanation of the effect. Let E^0 and $\overrightarrow{P^0}$ be the energy and the momentum of an incident photon, ϵ^0 and $\overrightarrow{p^0}$ energy and momentum of the electron before the collision, and let E' and $\overrightarrow{P'}$, ϵ' and $\overrightarrow{p'}$ stand for the corresponding values after the collision. These values are then connected by the conservation rules

$$E^0+\epsilon^0=E'+\epsilon', \quad \overrightarrow{P^0}+\overrightarrow{p^0}=\overrightarrow{P'}+\overrightarrow{p'}. \tag{27}$$

In addition there are general relations between energy and momentum for photons and free electrons:

$$E=P\,.\,c, \quad \epsilon=\frac{1}{2m}p^2=\text{kinetic energy}.$$

For given initial values $\overrightarrow{P^0}$ and $\overrightarrow{p^0}$ and for a given *direction* of the deflected photon (27) determines $\overrightarrow{P'}$, E', $\overrightarrow{p'}$, ϵ'. In particular we may take $\epsilon^0=0$ and $p^0=0$ (electrons at rest originally) and obtain then for any given direction the dependence of the secondary light frequency $\nu'=E'/h$ on $\nu^0=E^0/h$ and on the mass m of the electron. This corpuscular explanation explains not only the dependence of the frequency of the secondary light on the direction, but accounts also for the observations of Bothe and Geiger that there are coincidences in time between individual deflected electrons and deflected photons, which according to Compton and Simon satisfy the conservation rules in every single instance, not only in the average.

Schrödinger has explained the same facts with the help of the *wave theory*. We must first form the "transition density" of the electronic matter (see (23′) and (34) of Part I)

$$(28) \qquad \chi(\epsilon^0, p^0) \,.\, \chi^*(\epsilon', p') \,.\, e^{\frac{2\pi i}{h}[(p^0 - p').r + (\epsilon^0 - \epsilon')t]}.$$

(r and p are vectors.) If we use unit vectors s^0 and s' parallel to p^0 and p', and introduce periodicities λ and ν in space and time by

$$(28') \qquad p^0 = \frac{h}{\lambda^0} s^0, \quad \epsilon^0 = h\nu^0, \quad \ldots,$$

then we obtain the transition density in the form

$$(29) \quad \rho(r,t) = \text{const. exp. } 2i\pi\left[\left(\frac{s^0}{\lambda^0} - \frac{s'}{\lambda'}\right).r + (\nu^0 - \nu')t + \text{const.}\right].$$

ρ represents the intensity of a plane beat wave, the superposition of the incident and the reflected waves of matter. The transition density $\rho(r,t)$ serves then as a Huygens source of secondary light waves emitted under the influence of the primary incident light waves. Introducing two plane light waves characterized by their wave length Λ, frequency N and unit vector S of direction, we obtain a superposed light field with the *beat* intensity

$$(29') \quad I(r,t) = \text{const. exp. } 2i\pi\left[\left(\frac{S'}{\Lambda'} - \frac{S^0}{\Lambda^0}\right).r + (N' - N^0)t + \text{const.}\right].$$

This light field will be in equilibrium with the matter density $\rho(r,t)$ only if at every point r and at every time t both ρ and I have the same phase (or at least the same constant phase difference).

This means, however, that the coefficients of r and t in both expressions must be the same:

$$(30) \qquad \frac{s^0}{\lambda^0}-\frac{s'}{\lambda'}=\frac{S'}{\Lambda'}-\frac{S^0}{\Lambda^0}, \quad \nu^0-\nu'=N'-N^0.$$

By virtue of (28′), these relations are identical with the conservation rules of (27). So the corpuscular description (conservation rules) and the wave description (phase relation) account for the same dependence of the secondary colour on the angle of deflection. In order to explain the variation of *intensity* with the direction we need a more detailed theory of the reaction between matter and light, the result being dependent on the ratio e/m.

§ 30. BOTHE-GEIGER AND COMPTON-SIMON EXPERIMENTS

The apparent simultaneity of the deflected light and the recoiling matter as observed by Bothe and Geiger (14) is an immediate consequence of the corpuscular theory of collisions between photons and *particles* of matter. In order to explain it from the complementary standpoint of *waves* we must confine the transition density ρ of (29) to a volume ΔV according to the dimensions of the real apparatus, and to a small time interval Δt equal to the time interval between two successive readings. ρ of (29) then represents a *finite* crystal with travelling periodicities which acts only for a short time Δt as a Huygens source and diffracts the incident light waves towards a certain direction α and its surrounding $\Delta\alpha$ producing a Laue spot. The width $\Delta\alpha$ increases with decreasing size ΔV of the "crystal", and the range $\Delta\nu'$ of the diffracted frequency increasing with decreasing observation time Δt. If ΔV is not too small, the diffracted cone $\Delta\alpha$ emerging from the finite crystal ΔV will appear as a corpuscular-like "needle radiation".

In order to explain that the transition density ρ is limited to intervals ΔV and Δt only, one may consider two wave packets of matter limited in space and moving straight ahead so that they overlap in ΔV during Δt only (Fig. 11). Since the transition density ρ is the *product* of the contributions of each wave packet (cf. (28)), ρ will be different from zero and will send out a secondary light cone $\Delta\alpha$ only during the time and at the place of their overlapping. The time coincidences of Compton and Simon (15) are thus explained within the accuracy allowed for waves by the rules of the resolving power or allowed for particles by the uncertainty principle.

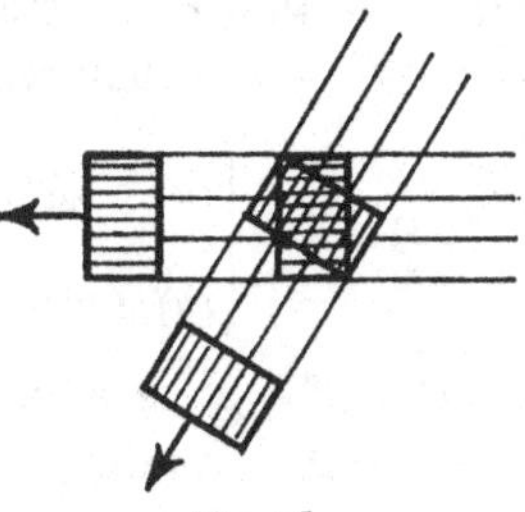

Fig. 11.

§ 31. DOPPLER EFFECT; RAMAN EFFECT

The Doppler effect has always been considered as a conclusive argument in favour of the undulatory theory of light: Waves emitted by a vibrator of frequency N moving with the velocity v and viewed from the direction of ϕ (Fig. 12) display a frequency $N+\Delta N$ where

$$\frac{\Delta N}{N}=\frac{v}{c}\cos\phi. \tag{31}$$

In order to explain the Doppler shift by means of the corpuscular theory, one has to suppose an atom to emit a photon of

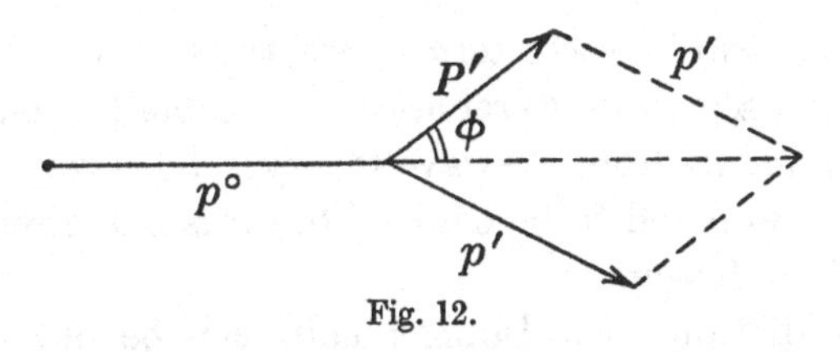

Fig. 12.

the energy E' and of the momentum P' during its transition from a state ϵ^0 and p^0 to ϵ' and p'. The conservation rules then demand that

$$\epsilon^0-\epsilon'=E' \quad \text{and} \quad p^0-p'=P'. \tag{32}$$

The energy of the atom may consist of its kinetic energy $(1/2m)p^2$ plus an intrinsic energy U. The energy of the photon is $E' = cP'$. (32) then reads (refer to Fig. 12)

$$(33) \quad \begin{cases} (a) \quad E' = U^0 + \frac{1}{2m}(p^0)^2 - U' - \frac{1}{2m}(p')^2, \\ (b) \quad (p')^2 = (p^0)^2 + (P')^2 - 2p^0 . P' \cos\phi. \end{cases}$$

If $P' = p^0 - p'$ is small compared with $p^0 + p'$, we may write $p^0 - p' = \Delta p$ and obtain approximately

$$(34) \quad \begin{cases} (a) \quad E' - (U^0 - U') = \frac{1}{m} p^0 . \Delta p, \\ (b) \quad (p^0 - \Delta p)^2 = (p^0)^2 + (\Delta p)^2 - 2p^0 \frac{E'}{c} \cos\phi. \end{cases}$$

Since $U^0 - U'$ would be the energy of the photon if emitted by the atom at rest, $E' - (U^0 - U') = \Delta E$ represents the excess energy of the photon emitted by the atom in motion over the energy E emitted by the atom at rest. So we obtain, neglecting small terms of the second order:

$$(35a) \quad \Delta E = \frac{1}{m} p \Delta p,$$

$$(35b) \quad -2p\Delta p = -2p \frac{E}{c} \cos\phi.$$

Eliminating Δp from the last two equations, we obtain

$$(36) \quad \frac{\Delta E}{E} = \frac{p}{mc} \cos\phi = \frac{v}{c} \cos\phi.$$

Expressing (36) in wave terms with the help of $E = hN$ and $\Delta E = h\Delta N$, we see that (36) represents the same Doppler shift as (31). The difference from the wave theory is, however, that single photons are supposed to be emitted towards one direction at a time (needle radiation).

Needle radiation with Doppler shift can be obtained with the help of the complementary "transition density" serving as Huygens sources of secondary light waves. The only difference from the Compton effect is that here no incident light is present, E^0 and P^0 being zero in the formulas of the preceding paragraph.

The transition of a particle of matter $\epsilon^0 \rightarrow \epsilon'$, $p^0 \rightarrow p'$ corresponds now to the "transition density" $\rho(r, t)$ given in (29). The light field that is in equilibrium with this periodic density (crystal) must coincide in its phase at every point and at every time with the phase of ρ. Since in this case only one light wavé N', Λ', S' is present, (30) reduces to

$$\nu^0 - \nu' = N', \quad \frac{s^0}{\lambda^0} - \frac{s'}{\lambda'} = \frac{S'}{\Lambda'}. \tag{37}$$

These equations (37), if multiplied by h, give again the conservation rules (32).

Half-way between the Compton and the Doppler effect is the Raman effect, concerning the changed colour of secondary waves emerging from atoms that are irradiated with monochromatic primary light and that are in a state of transition of their intrinsic energy from U^0 to U'. Again one can account for the observed colour shift either by corpuscular transition processes with conservation of energy, or by a corresponding transition density as the Huygens source of secondary waves.

In all of these cases a more specialized theory of the reaction of charged matter with light is needed if one wishes to explain the distribution of the intensity at various angles of observation.

§ 32. ELEMENTARY BUNDLES OF RAYS

Waves of light or matter starting from a line element δx necessarily spread over a bundle $\delta\alpha$ of directions: According to the elementary theory of diffraction (6) we have

$$\delta\alpha = \frac{\lambda}{\delta x \sin\alpha}.$$

Since $\sin\alpha\,\delta\alpha = \delta(\cos\alpha)$, we obtain the fundamental relations for the *optical* resolving power

$$\frac{\delta x . \delta(\cos\alpha)}{\lambda} = 1 \quad \text{and} \quad \frac{\delta y . \delta(\cos\beta)}{\lambda} = 1. \tag{38}$$

Furthermore, if a wave train has the finite length δl, then its Fourier analysis shows it to consist at least of a range of different wave lengths λ or wave numbers $n=1/\lambda$ in an interval

$$\delta n=\delta\left(\frac{1}{\lambda}\right)=\frac{1}{\delta l}.$$

Thus we obtain the equation of the *spectral* resolving power:

$$(38') \qquad \delta\left(\frac{1}{\lambda}\right).\delta l=1.$$

M. Laue (16) has introduced the concept of an "elementary bundle" of waves, that is, a bundle of such a focal area $\delta x.\delta y$, angular aperture $\delta\Omega=\dfrac{\delta(\cos\alpha)\,\delta(\cos\beta)}{\cos\gamma}$, longitudinal length δl and range of wave number $\delta(1/\lambda)$, that (38) and (38′) are satisfied simultaneously; that is

$$(39) \qquad \frac{\delta x.\delta(\cos\alpha)}{\lambda}.\frac{\delta y.\delta(\cos\beta)}{\lambda}.\delta l.\delta\left(\frac{1}{\lambda}\right)=1.$$

We now translate this definition of an elementary bundle into the language of *corpuscles*. If p is their absolute momentum, then the interval of directions $\delta(\cos\alpha)$ and $\delta(\cos\beta)$ implies an interval of components

$$\delta p_x=p.\delta(\cos\alpha) \quad\text{and}\quad \delta p_y=p.\delta(\cos\beta).$$

According to de Broglie's relation $p=h/\lambda$, we can translate the interval $\delta(1/\lambda)$ of wave numbers into an interval $\delta p=h.\delta(1/\lambda)$ of momenta. So we obtain from (38) and (38′) the boundaries of an elementary bundle determined by the corpuscular equations

$$\delta x.\delta p_x=h,\quad \delta y.\delta p_y=h,\quad \delta l.\delta p=h,$$

their product being h^3. In the first two relations we recognize the elementary intervals of uncertainty of position and momentum. The last relation $\delta l.\delta p=h$ is identical with the uncertainty relation $\delta\epsilon.\delta t=h$ of energy and time. We may prove this for matter and for light if we multiply the following relations:

$$\text{matter:}\begin{cases}\delta t=\dfrac{\delta l}{v}=\dfrac{m}{p}\delta l,\\[2ex] \delta\epsilon=\delta\left(\tfrac{1}{2}mv^2\right)=\dfrac{p}{m}\delta p.\end{cases} \qquad \text{light:}\begin{cases}\delta t=\dfrac{\delta l}{c},\\[2ex] \delta\epsilon=c\,\delta p.\end{cases}$$

An elementary bundle of corpuscles can thus be characterized as satisfying the relation

$$(39') \qquad \frac{\delta x \,.\, \delta p_x}{h} \,.\, \frac{\delta y \,.\, \delta p_y}{h} \,.\, \frac{\delta t \,.\, \delta \epsilon}{h} = 1.$$

It constitutes a kind of smallest unit. Parts of it, for instance bundles diverging from an opening δx into an angular range smaller than $\delta\alpha = \frac{\lambda}{\delta x \,.\, \sin\alpha}$, can neither be produced nor observed.

§ 33. JEANS' NUMBER OF DEGREES OF FREEDOM

We may now ask with Jeans (17) what is the number of elementary bundles belonging to a larger range Δx, $\Delta(\cos\alpha)$, ...? The answer is given by the number

$$(40) \qquad \Delta Z = \frac{\Delta x \,.\, \Delta(\cos\alpha)}{\lambda} \,.\, \frac{\Delta y \,.\, \Delta(\cos\beta)}{\lambda} \,.\, \Delta l \,.\, \Delta\left(\frac{1}{\lambda}\right),$$

or in corpuscular language by

$$(40') \qquad \Delta Z = \frac{\Delta x \,.\, \Delta p_x}{h} \,.\, \frac{\Delta y \,.\, \Delta p_y}{h} \,.\, \frac{\Delta l \,.\, \Delta p}{h}.$$

Next we ask how many elementary bundles, belonging to an interval $\Delta(1/\lambda)$ or Δp, are found in a certain volume ΔV? We can write

$$\Delta x \Delta y = \Delta f, \quad \Delta\cos\alpha \,.\, \Delta\cos\beta = \cos\gamma \,.\, \Delta\Omega, \quad \Delta f \,.\, \Delta l \cos\gamma = \Delta V.$$

Then, considering bundles in all directions, we must replace $\Delta\Omega$ by 4π. So we find the number of elementary bundles

$$(41) \qquad \Delta Z = \Delta V \,.\, 4\pi \frac{1}{\lambda^2} \Delta\left(\frac{1}{\lambda}\right) \quad \text{(waves)},$$

$$(41') \qquad \Delta Z = \Delta V \,.\, 4\pi \frac{p^2 \Delta p}{h^3} \quad \text{(corpuscles)}.$$

Jeans' number ΔZ is also the number of degrees of freedom in ΔV for the interval $\Delta(1/\lambda)$ or Δp. That is, in order to describe the physical state in ΔV for this interval $\Delta(1/\lambda)$ or Δp we must determine all the intensities of the ΔZ elementary bundles that are present in the volume. Some of them may have a large intensity, some may have zero intensity. If the rays are polarizable we must add a factor 2 on the right-hand sides of (41) and (41′). For example, we may ask for the number of visible "sunbeams" (elementary bundles) per cubic centimetre. Here we have $\Delta V = 1$ and λ from 4 to 7×10^{-5} cm. with $\Delta(1/\lambda) = (\frac{1}{4} - \frac{1}{7})\, 10^5$, hence

$$\Delta Z \sim 10^{14} \text{ visible elementary beams in } V = 1.$$

The number of elementary bundles (equal to the number of degrees of freedom) is of great importance for the thermodynamic and statistical properties of matter or light enclosed in a volume.

§ 34. UNCERTAINTY OF ELECTROMAGNETIC FIELD COMPONENTS

We must not be misled by thinking that the theory of wave functions ψ and χ represents a wave theory of matter. A real wave theory of matter ought to suppose that we can measure, at least in principle, the amplitudes and phases of the waves themselves. The complex functions $\psi(q)$ and $\chi(p)$, however, are unobservable in principle. For even if both real functions $\rho(q)$ and $\sigma(p)$ are given or measured, the *phases* of ψ and χ contain arbitrary additive constants that have no physical meaning whatsoever. χ and ψ are functions characteristic for quantum theory, not for a real wave theory of matter.

One should be able, however, to develop, as a counterpart to the classical corpuscular theory, a classical wave theory of matter with real amplitudes and phases assumed to be measurable quantities. Such a theory has been developed by D. R. Hartree (18), and it turned out that his "classical" wave theory explains all the macroscopic properties of matter just as well as does the classical corpuscular theory. But it fails within microscopic limits, because of the uncertainty rules of the wave amplitudes and phases.

Instead of giving here a report of Hartree's wave theory of matter, let us consider another example of a classical wave theory, the Maxwell theory of the electromagnetic field. Let us ask then within what limits this theory is applicable, and where it is to be modified by quantum theory.

Maxwell's theory has to be limited in its application in such a way as not to lead into a conflict with the apparent corpuscular structure of light. If we observe, for example, the electromagnetic momentum p in a certain volume element $(\Delta q)^3$ in the field of light waves λ, this momentum cannot be told more accurately than up to the momentum h/λ of one photon. The smallest photon in $(\Delta q)^3$ is that of wave length $\lambda = \Delta q$ and of momentum $h/\Delta q$. Thus the electromagnetic momentum in $(\Delta q)^3$ has at least the uncertainty $\delta p = h/\Delta q$.

On the other hand, if E and H are the averages of the field vectors in $\Delta V = (\Delta q)^3$, then

$$p = \Delta V \frac{1}{4\pi c} E \times H$$

represents the electromagnetic momentum in ΔV. If δE and δH are margins of uncertainty for E and H, then we have as the uncertainty of p

$$\begin{aligned}\delta p &= \frac{\Delta V}{4\pi c}\{(E+\delta E) \times (H+\delta H) - E \times H\} \\ &= \frac{(\Delta q)^3}{4\pi c}\{E \times (H+\delta H) + (E+\delta E) \times H + \delta E \times \delta H\}.\end{aligned}$$

Even in the case of $E = 0$ and $H = 0$ there remains a minimum of uncertainty

$$\delta p = \frac{(\Delta q)^3}{4\pi c} \delta E \times \delta H.$$

This equation, together with $\delta p = h/\Delta q$, leads to the uncertainty relation for the electromagnetic field components:

$$\delta E \times \delta H = \frac{4\pi h c}{(\Delta q)^4}.$$

The vectorial cross-product $\delta E \times \delta H$ concerns the pairs of perpendicular components, for instance,

$$\delta E_x \times \delta H_y = \frac{4\pi h c}{(\Delta q)^4}.$$

The result is: Two perpendicular components like E_x and H_y in $\Delta V = (\Delta q)^3$ cannot be measured simultaneously in the same volume element better than within the uncertainties δE_x and δH_y whose product is $4\pi hc/(\Delta q)^4$. There are no restrictions however on the measurement of E_x in one volume element and H_y in another, or on the measurement of two parallel components E_x and H_x in the same volume element.

In conclusion to Part II we may say that quantum theory is based on the following fundamental fact. A quantity of free matter which is condensed at $t=0$ to a range δx_0 is bound to spread during $\delta t'$ over a range

$$\delta x' = \frac{h}{m} \cdot \frac{\delta t'}{\delta x_0}. \quad \text{(Compare with (22).)}$$

This spread may be explained as resulting from an original spread of the corpuscular velocities $\delta V_0 = h/m . \delta x_0$ or momenta $\delta p_0 = h/\delta x_0$, if we assume that the matter consists of mechanical particles of mass m.

The complementary consideration would be as follows. It is a basic fact that a quantity of matter, originally able to give out momenta within range δp_0 (as seen from the diffraction of photons), cannot be made to decrease this range to less than

$$\delta p' = \frac{hm}{\delta t' . \delta p_0} \qquad \begin{pmatrix} \text{where } \delta p_0 = h/\delta x_0 \\ \text{and } \quad \delta p' = h/\delta x' \end{pmatrix}$$

after the time $\delta t'$. This ever-decreasing range can be explained by an ever-decreasing range of the *group velocities* of the *waves* that constitute the matter. Its wave numbers $N_0 = 1/\lambda_0$ spread over $\delta N_0 = 1/\delta x_0 = h/\delta p_0$ and its frequencies over $\delta\nu = 1/\delta t'$, so its group velocities $\delta\nu/\delta N_0$ spread over $\delta g' = \dfrac{h}{\delta t' . \delta p_0}$. In order to explain the former formula for $\delta p'$ one needs the relation

$$\delta p' = m . \delta g'.$$

In other words, one has to assume that a packet of matter waves containing a range $\delta g'$ of group velocities is able to give out momenta in the range $\delta p' = m . \delta p'$, where m appears now as a characteristic constant pertaining to the substance of the waves.

So we see that it is incorrect to consider the existence of characteristic constants m as an argument in favour of the corpuscular structure of matter.

PART III

THE PRINCIPLE OF INTERFERENCE AND SCHRÖDINGER'S EQUATION

§ 35. PHYSICAL FUNCTIONS

Let $Q(q,p)$ be a real function of the $2N$ quantities $q_1 q_2 \ldots q_N$ and $p_1 p_2 \ldots p_N$. If the q's and p's are supposed to mean co-ordinates and momenta [meaning stands for certain prescriptions for measuring them], then $Q(q,p)$ means a certain physical quantity also. This quantity will have, in general, no name. Only the most important physical functions have received names; for instance, those obeying rules of conservation. If the q's and p's vary in time according to the equations

$$\dot{q}_K = \frac{\partial Q}{\partial p_K}, \quad \dot{p}_K = -\frac{\partial Q}{\partial q_K},$$

then Q will be constant in time and Q is called the energy. The formof the function $Q(q,p)$ is in no way significant for its meaning. A certain function Q, say $Q = \frac{1}{2m_0} p^2 + \frac{k}{2} q^2$, may represent the energy of a Newtonian mass point m_0 under the quasi-elastic force kq. But if we have in mind a relativistic particle with a variable mass m, and m_0 stands for the rest mass only, then the same function Q no longer has the meaning of energy, although it still has some physical meaning.

These remarks are to emphasize that it is only a matter of interpretation as to what is the physical meaning of a mathematical function $Q(q,p)$. It depends on the meaning of the symbols q and p (rectangular or curvilinear co-ordinates) and the "kind" of particle we have in mind (the equations of motion applying to q and p). But irrespective of such questions of interpretation we may solve mathematical problems involving a function $Q(q,p)$. For example, we can solve the above equations of motion with Q as an "energy function". And we can ask for a mathematical method to determine the function $\sigma(Q)$ which shall mean the abundance of various values of the function Q in a certain experimental set-up, supposing that the "set-up" and the "abundance" are defined mathematically by giving certain values to certain other functions of q and p and introducing certain prescriptions for connecting them with σ.

§ 36. INTERFERENCE OF PROBABILITIES FOR p AND q

The most important result of Part I was the development of a definite relation between the density $\rho(q)$ of free particles in space and the abundance $\sigma(p)$ of various momenta in a certain physical set-up of matter sources, shutters, eyepieces and the like. The relation between ρ and σ was expressed in terms of the amplitudes $\psi(q)$ and $\chi(p)$ by the formulae (16) of Part I:

$$\psi(q) = \int \chi(p) e^{\frac{2\pi i}{h} p.q} dp \tag{1}$$

and its inverse†

$$\chi(p) = \frac{1}{h} \int \psi(q) e^{-\frac{2\pi i}{h} q.p} dq. \tag{1'}$$

It is very significant that the amplitudes ψ and χ appear as harmonic expansions, as Fourier series or integrals, of each other, a fact intimately connected with our observation of ρ and σ by optical means.

To put this Fourier relation in evidence, we may write ψ and χ in the form

$$\psi(q) = \int \chi(p) \Psi_p(q) dp, \tag{2}$$

$$\chi(p) = \int \psi(q) X_q(p) dq, \tag{2'}$$

† In the case of a system with N co-ordinates q_K and momenta p_K one has to replace $p.q$ by $\Sigma p_K q_K$ and integrate over $dq_1, dq_2, dq_3, \dots$ instead of over dq, and over $dp_1, dp_2, dp_3, \dots$ instead of over dp. In order to eliminate the factor $1/h$ or $1/h^N$ before the integral in (1′) we may introduce units of q and p whose product is h.

introducing the standard wave functions

$$(2'') \qquad \Psi_p(q) = e^{\frac{2\pi i}{h} p.q} \quad \text{and} \quad X_q(p) = e^{-\frac{2\pi i}{h} q.p}.$$

$\rho(q) = |\psi(q)|^2$ has been interpreted to mean the probability of finding a particle (not a certain individual particle but any particle at all) at the point† q and $\sigma(p) = |\chi(p)|^2$ as the probability of finding a particle of the momentum p. In our set-up $\psi(q)$ and $\chi(p)$ are the corresponding *probability amplitudes* for q and p.

If the corpuscular picture were correct, we should be able to compose $\rho(q)$ and $\sigma(p)$ by the ordinary product rules for composing probabilities:

$$\rho(q) = \int \sigma(p) \,.\, U_p(q)\, dp,$$

$$\sigma(p) = \int \rho(q) \,.\, V_q(p)\, dq,$$

where $U_p(q)$ would mean the standard probability, independent of any set-up, that a particle whose momentum is p be found at q. And $V_q(p)$ would mean the probability that a particle whose position is q be found at p. Instead of these probability rules we have the relations (2) and (2′) between the probability *amplitudes* $\psi(q)$ and $\chi(p)$ and the "amplitudes" (2″) of the two probabilities $U_p(q)$ and $V_q(p)$ defined before. The equations (2) and (2′) present the rules of *interference of probabilities* which replace the classical probability rules of the last two equations.‡

† More correctly: In an interval $\delta q = 1$ around q. We may be allowed to omit the reference to these unit intervals.

‡ Analogous relations hold for energy and time:

$$\psi(t) = \int \chi(E)\, e^{(2\pi i/h)\,.\,E\,.\,t}\, dE$$

and its inverse $$\chi(E) = \frac{1}{h} \int \psi(t)\, e^{-(2\pi i/h)\,.\,E\,.\,t}\, dt.$$

Introducing standard amplitudes

$$\Psi_E(t) = e^{(2\pi i/h)\,.\,E\,.\,t} \quad \text{and} \quad X_t(E) = e^{-(2\pi i/h)\,.\,E\,.\,t},$$

we obtain the interference formulae

$$\psi(t) = \int \chi(E)\, \Psi_E(t)\, dE \quad \text{and} \quad \chi(E) = \int \psi(t)\, X_t(E)\, dt.$$

We can see in the interference rules (2) and (2′) the most spectacular expression of the failure of the corpuscular theory. Nevertheless, it is still very convenient to use corpuscular terms, at least in this mathematical Part, without being misunderstood as to the physical meaning of these terms. Using q' for a fixed value of q, the function $X_{q'}(p)$ of (2″) is the probability amplitude for a single particle which is in a definite position $q=q'$ to possess the momentum p at the same time. One can say also that $X_{q'}(p)$ is the amplitude of the probability of finding a momentum p in the "pure case" or standard set-up, in which all particles have the same position $q=q'$. A similar interpretation applies to $\Psi_{p'}(q)$. According to their physical meaning, the absolute squares of the *standard* functions $X_q(p)$ and $\Psi_p(q)$ of (2″) are equal. The amplitudes $\psi(q)$ and $\chi(p)$ in (2) and (2′), however, refer to a special set-up that represents in general a "mixed case", with various p and q values being present at the same time.

§ 37. GENERAL INTERFERENCE OF PROBABILITIES

We now come to the central problem of quantum mechanics. Let $Q(q,p)$ be any physical quantity defined as a *real* function of q and p. Is it possible then to predict the abundance $\tau(Q)$ of various values of $Q(q,p)$ in a certain set-up which may be characterized by a given density amplitude $\psi(q)$ or a given abundance amplitude $\chi(p)$? This question is answered in quantum mechanics in three successive steps:

First, consider the abundance $\tau(Q)$ as the absolute square of an abundance amplitude $\phi(Q)$.

Second, express $\phi(Q)$ in terms of $\psi(q)$ or $\chi(p)$ by the following integrals that are similar to the interference rules (2) and (2′):

(3) $$\phi(Q)=\int \chi(p)\,.\,\Phi_p(Q)\,dp,$$

and also

(3′) $$\phi(Q)=\int \psi(q)\,.\,\Psi_q(Q)\,dq.$$

Here $\Phi_p(Q)$ and $\Psi_q(Q)$ are certain *standard* amplitudes belonging to "pure cases": For instance $\Phi_{p'}(Q)$ means (in corpuscular terms) the probability amplitude that a particle, which has the momentum $p=p'$, possesses at the same time the value Q of the function $Q(q,p)$. Or in other words $|\Phi_{p'}(Q)|^2$ means the relative number of particles possessing the value Q of $Q(q,p)$ in a standard set-up in which all particles have the same value p' of p.

Third, calculate the standard amplitudes such as $\Phi_p(Q)$ and $\Psi_q(Q)$. Quantum mechanics has developed the proper mathematical method for doing this (§ 40). But even before learning about this method we can immediately deduce an important general relation between various standard amplitude functions. Our set-up, characterized by $\psi(q)$ and $\chi(p)$, may represent a particular "pure case" in which all particles have the same value β' of a certain other physical quantity $\beta(q,p)$; we may write in this particular case $\chi_{\beta'}(p)$ and $\psi_{\beta'}(q)$ for $\chi(p)$ and $\psi(q)$. Then (3) and (3′) take on the special forms

$$\phi_{\beta'}(Q)=\int\chi_{\beta'}(p).\Phi_p(Q)\,dp, \tag{4}$$

$$\phi_{\beta'}(Q)=\int\psi_{\beta'}(q).\Psi_q(Q)\,dq. \tag{4′}$$

Since $\beta(q,p)$ and $Q(q,p)$ may be any real functions of q and p, the last equations represent a great variety of cases. In a more schematic way we may write (4) and (4′) in the form of the *general theorem of interference of probabilities*:

$$F_{A'}(C')=\int G_{A'}(B').H_{B'}(C')\,dB', \tag{5}$$

where A', B', and C' are values of any physical quantities A,B,C which may be defined as functions of p and q. The absolute squares of the two probability amplitudes $\Phi_\beta(Q)$ and $\Psi_Q(\beta)$ must be equal, since *both* express the real probability of finding the value A of a physical function $A(q,p)$ and the value B of another function $B(q,p)$ simultaneously:

$$|\Phi_\beta(Q)|^2=|\Psi_Q(\beta)|^2. \tag{5′}$$

The amplitudes $\Phi_\beta(Q)$ and $\Psi_Q(\beta)$ themselves, however, may still differ by a complex factor $e^{i\alpha}$ of the absolute value 1, and there is no possibility of observing the phase α. So the choice of the phase is free, provided that (5) is not violated. In § 47 we shall see that we have to choose the phase so that

$$\Phi_\beta(Q) = \Psi_Q^*(\beta); \quad \text{hence} \quad \Phi_\beta^*(Q) = \Psi_Q(\beta), \tag{6}$$

where the * indicates the complex conjugate. That is, the standard probability amplitudes assume their complex conjugate values if their arguments are exchanged with their lower indices. In (2″) we already had an example of $\Psi_p(q)$ being the complex conjugate of $X_q(p)$.

§ 38. DIFFERENTIAL EQUATIONS FOR $\Psi_p(q)$ AND $X_q(p)$

We must now determine the general method of calculating standard functions like $\Phi_p(Q)$, $\Psi_q(Q)$, $\phi_\beta(Q)$. The method can be developed as a generalization of the already known case (2″). Note that the standard function (2″)

$$\Psi_{p'}(q) = e^{\frac{2\pi i}{h}p'q}$$

is a solution of the following differential equation:

$$\frac{h}{2\pi i}\frac{\partial}{\partial q}\Psi = p'\Psi. \tag{7}$$

This equation "corresponds" to the classical equation:

$$p = p'. \tag{7'}$$

The correspondence can be made more evident by writing the differential equation (7) in the symbolic form

$$\mathbf{p}\psi = p'\psi, \tag{7''}$$

where the symbol $\mathbf{p}$ is equivalent to $\frac{h}{2\pi i}\frac{\partial}{\partial q}$. In the case of several co-ordinates and momenta we introduce the symbol

$$\mathbf{p}_K = \frac{h}{2\pi i}\frac{\partial}{\partial q_K}. \tag{8}$$

The probability amplitude (cf. (16′), Part I)

$$(8') \qquad e^{\frac{2\pi i}{h}(p_1' q_1 + p_2' q_2 + \dots)} = \psi_{p_1' p_2' \dots}(q_1 q_2 \dots),$$

called $\psi_{p'}(q)$ in short, is then the solution of the set of simultaneous differential equations:

$$(8'') \qquad \mathbf{p}_1 \psi = p_1' \psi, \ \mathbf{p}_2 \psi = p_2' \psi, \ \dots,$$

which corresponds to the set of classical equations

$$p_1 = p_1', \ p_2 = p_2', \ \dots.$$

Similarly, the standard function $X_{q'}(p) = e^{-\frac{2\pi i}{h} p q'}$ of (2″) is the solution of the differential equation

$$\mathbf{q} X = q' X,$$

where we have introduced the symbol

$$\mathbf{q} = -\frac{h}{2\pi i} \frac{\partial}{\partial p}.$$

The minus sign on the right in contrast with the plus sign in (8) corresponds to the well-known fact that if p is considered as a "co-ordinate", then $-q$ is its conjugate momentum.

§ 39. DIFFERENTIAL EQUATION FOR $\phi_\beta(q)$

We must now generalize this method. Here we follow the basic investigations of M. Born, P. Jordan, and F. London. Let $\beta_L(q, p)$ with $L = 1, 2, \dots$ be a set of real functions representing a set of physical quantities. The number of these functions may be *equal*† to the number of co-ordinates q_K. Let us consider the "pure case" in which each particle in our physical set-up has the values

$$(9) \qquad \beta_L(q, p) = \beta_L' \quad (L = 1, 2, \dots).$$

We then wish to determine the amplitude $\phi_{\beta'}(q)$ of the probability of finding a particle at q. As a generalization of (8″) quantum mechanics suppose $\phi_{\beta'}(q)$ to be the solution ("eigen-function") of the following set of differential equations:

$$(9') \qquad \beta_L(q, \mathbf{p}) \phi_{\beta'}(q) = \beta_L' \cdot \phi_{\beta'}(q) \quad (L = 1, 2, \dots).$$

† If the number of functions β_L is smaller than the number of the q_K's, then one can introduce some of the q_K's themselves as additional functions $\beta_L = q_L$ until the number of β's is *equal* to the number of the q_K's.

The operators $\beta_L(q, \mathbf{p})$ in (9′) mean that we must replace† the momenta p in the functions $\beta_L(q, p)$ by the differential operators (8). On the right-hand side we have a simple product of the value β'_L with ϕ. It is the outstanding feature of these differential equations, that they sometimes possess unique and finite solutions $\phi_{\beta'}(q)$ only for certain selected values β'_L, the so-called eigen-values of the differential equations.

If $Q_M(q, p)$ be another set of physical functions, their number being equal to the number of co-ordinates q_K, and we consider the "pure case" that each particle possesses the values

$$(10) \qquad Q_M(q, p) = Q'_M \quad (M = 1, 2, \ldots),$$

then $\Psi_{Q'}(q)$ is to be calculated similarly as the solution (eigen-function) of the set of differential equations:

$$(10') \qquad Q_M(q, \mathbf{p})\, \Psi_{Q'}(q) = Q'_M \,.\, \Psi_{Q'}(q) \quad (M = 1, 2, \ldots).$$

These equations again have sometimes only certain selected eigen-values Q' for which $\Psi_{Q'}(q)$ is finite and unique.

The final test of this method of calculating standard amplitudes $\phi_{\beta'}(q)$ and $\Psi_{Q'}(q)$ is found in its agreement with the observations. Mathematically the method represents the natural generalization of the differential equations (8″) for the wave functions (8′) which were inferred from the experiments reported in Part I.

The new functions $Q_M(q, p)$ or $\beta_L(q, p)$ can be considered as two sets of "new co-ordinates".

§ 40. THE GENERAL PROBABILITY AMPLITUDE $\Phi_{\beta'}(Q)$

A new and more general problem arises if we demand the amplitude $\Phi_{\beta'}(Q)$ of the probability of finding a particle with certain values Q_M of $Q_M(q, p)$ when the same particle possesses the values β'_L of $\beta_L(q, p)$. There are two methods of finding $\Phi_{\beta'}(Q)$. One

† For instance, if $\beta(q, p)$ is the function $(q^2 + p^2)$, then $\beta(q, \mathbf{p})$ is the operator $\left\{q^2 + \left(\frac{h}{2\pi i}\right)^2 \frac{\partial^2}{\partial q^2}\right\}$, and the left-hand side of (9′) reads $q^2 \,.\, \phi - \frac{h^2}{4\pi^2} \frac{\partial^2 \phi}{\partial q^2}$.

method is offered by the general interference formula (5), which in the present case reads:

$$(11)\qquad \Phi_{\beta'}(Q)=\int \phi_{\beta'}(q)\,.\,\Psi_q(Q)\,dq \equiv \int \phi_{\beta'}(q)\,.\,\Psi_Q^*(q)\,dq,$$

where $\phi_{\beta'}(q)$ is the solution (eigen-function) of (9′), and

$$(12)\qquad \Psi_Q^*(q)=\Psi_q(Q)\quad \text{(cf. (6))}$$

is the complex conjugate of the solution $\Psi_Q(q)$ of (10′).

The *other method* which leads to the same result† is to calculate the function $\Phi_{\beta'}(Q)$ directly as the solution of the differential equations

$$(13)\qquad \mathrm{B}_L(Q,\mathbf{P})\,\Phi_{\beta'}(Q)=\beta'_L\,.\,\Phi_{\beta'}(Q).$$

Here the operators $B_L(Q,\mathbf{P})$ are new forms of the operators $\beta_L(q,\mathbf{p})$, brought about by a *transformation* (Part V) of the original co-ordinates and differential operators

$$(13')\qquad q_K \text{ and } \mathbf{p}_K=\frac{h}{2\pi i}\frac{\partial}{\partial q_K}\quad\text{into}\quad Q_M \text{ and } \mathbf{P}_M=\frac{h}{2\pi i}\frac{\partial}{\partial Q_M}.$$

To give an example of such a transformation, we consider now a "point transformation".

§ 41. POINT TRANSFORMATIONS

The new co-ordinates Q_M may have the form $Q_M=Q_M(q)$, *not depending on the* p's. Then conversely we have

$$(14)\qquad q_K=q_K(Q).$$

In this case, called a point transformation, we have

$$(14')\qquad \mathbf{p}_K=\frac{h}{2\pi i}\frac{\partial}{\partial q_K}=\frac{h}{2\pi i}\sum_M \frac{\partial Q_M}{\partial q_K}\frac{\partial}{\partial Q_M}=\sum_M \frac{\partial Q_M}{\partial q_K}\mathbf{P}_M.$$

So we are led to a definite form of the operator $B_L(Q,\mathbf{P})$ in (13), namely

$$(14'')\qquad \beta_L(q,\mathbf{p})=\beta_L\left(q(Q),\sum_M \frac{\partial Q_M}{\partial q_K}\mathbf{P}_M\right)=\mathrm{B}_L(Q,\mathbf{P}).$$

† It is one of the basic theorems of quantum mechanics that the two methods (11) and (13) of calculating $\Phi_{\beta'}(Q)$ are identical; the proof is given in Part V. For a special example of confirmation see § 41.

We shall be faced later with the problem of finding the operators $\mathrm{B}_L(Q, \mathbf{P})$ in cases where the Q_M are given as functions of both the q and the p. In order to prepare now for this later problem we present the transition from $\beta_L(q, \mathbf{p})$ to $\mathrm{B}_L(Q, \mathbf{P})$ in a somewhat different form. We introduce the operator

$$S(q, \mathbf{P}) = \sum_M Q_M(q) \,.\, \mathbf{P}_M, \tag{15}$$

so that by a *formal* differentiation with respect to the symbol $\mathbf{P}_M$ one has

$$Q_M(q) = \frac{\partial S}{\partial \mathbf{P}_M}. \tag{15'}$$

If we now define

$$\mathbf{p}_K = \frac{\partial S}{\partial q_K}, \tag{15''}$$

this definition leads to the former expression (14′) by virtue of (15); and (14″) can be written in the simple form

$$\beta_L(q, \mathbf{p}) = \beta_L\left(q(Q), \frac{\partial S}{\partial q}\right) = \mathrm{B}_L(Q, \mathbf{P}).$$

In our present case of a point transformation, we can immediately verify the general theorem that (11) is equal to the solution of (13). *First*, we had in (9′)

$$\beta_L(q, \mathbf{p})\, \phi_{\beta'}(q) = \beta'_L \,.\, \phi_{\beta'}(q).$$

Second, instead of (10′), we have here simply

$$Q_M(q) \,.\, \Psi_{Q'}(q) = Q'_M \,.\, \Psi_{Q'}(q),$$

where $Q_M(q)$ is only a multiplier (not a differential operator, since it does not depend on $\mathbf{p}$). The last equation is solved by a function $\Psi_{Q'}(q)$ which vanishes everywhere except for the values $q_K = q_K(Q')$, where it possesses an infinitely steep maximum. *Third*, the point transformation $q_K = q_K(Q)$ says that the solution $\Phi_{\beta'}(Q)$ of (13) is only a transformed form of $\phi_{\beta'}(q)$, namely

$$\Phi_{\beta'}(Q) \equiv \phi_{\beta'}(q(Q)).$$

Consider now the interference integral (11):

$$\int \phi_{\beta'}(q)\, \Psi^*_{Q'}(q)\, dq.$$

Owing to the infinitely steep maximum of $\Psi^*_{Q'}(q)$ for $q=q(Q')$ this integral reduces to $\phi_{\beta'}(q(Q'))$, that is, to the value $\Phi_{\beta'}(Q')$. The identity of the solution of (13) with the interference integral (11) is thus verified in the case of a point transformation.

§ 42. GENERAL THEOREM OF INTERFERENCE

We must now find a general method for transforming the operators $\beta_L(q, \mathbf{p})$ into $\mathrm{B}_L(Q, \mathbf{P})$ in the case in which the Q_M are functions of both q and p. The general transformation method asked for must satisfy the following criterion. If the operators $\beta_L(q, \mathbf{p})$ are transformed into $\mathrm{B}_L(Q, \mathbf{P})$, then the eigen-function $\Phi_{\beta'}(Q)$ of (13) is again to be identical with the interference integral (11). In short we postulate the *general theorem of interference*:

A. *The eigen-functions of* (9′), (10′), (13) *satisfy the interference rule* (11).

§ 43. CONJUGATE VARIABLES

Without developing in detail the method of transformation itself (this will be done in Part V) we can determine an important criterion which it must satisfy. Consider the special case in which the transformed operators $B_L(Q, \mathbf{P}) = \beta_L(q, \mathbf{p})$ are the operators P_L themselves, and where $\beta'_L = P'_L$ are fixed values of the new momenta P_L. (13) reads in this special case

$$\mathbf{P}_L . \Phi_{P'}(Q) = P'_L . \Phi_{P'}(Q), \tag{16}$$

with

$$\mathbf{P}_L = \frac{h}{2\pi i} \frac{\partial}{\partial Q_L}.$$

This differential equation is solved by

$$\Phi_{P'}(Q) = e^{\frac{2\pi i}{h} P'.Q}. \tag{16′}$$

That is: the probability amplitude for a value Q_L, if the conjugate momentum has the value P_L, is of the same complex periodic form as was the probability amplitude in the original co-ordinates:

$$\phi_{p'}(q) = e^{\frac{2\pi i}{h} p'.q}.$$

Conversely, we can take this periodic form of the probability amplitude $\Phi_{P'}(Q)$ as a criterion for recognizing the conjugacy of co-ordinates and momenta $Q(q,p)$ and $P(q,p)$. As a consequence of (16′) we then obtain a Fourier relation between the amplitudes $F_{A'}(Q)$ and $G_{A'}(P)$ in any set-up A'

$$(17) \quad F_{A'}(Q) = \int G_{A'}(P) \, . \, \Phi_P(Q) \, dP = \int G_{A'}(P) \, . \, e^{\frac{2\pi i}{h} P.Q} \, dP,$$

as a special form of (5) and as a counterpart to (1).

§ 44. SCHRÖDINGER'S EQUATION FOR CONSERVATIVE SYSTEMS

If one† of the functions $\beta_{L'}$, say $\beta_N(q,p)$ in (9), represents the *energy* function $H(q,p)$, and $\beta'_N = E'$ is a fixed value of the energy, then the corresponding equation (9) reads

$$(18) \qquad H(q, \mathbf{p}) \, \psi_{E'}(q) = E' \, . \, \psi_{E'}(q).$$

This is the equation of Schrödinger, the most powerful mathematical instrument of the theory of atoms. Its importance is equal to the importance of the rule of conservation of energy in classical mechanics. The outstanding feature of this differential equation is that it has sometimes, depending on the special form of the energy function H, unique and finite eigen-functions only for certain selected eigen-values E'. Quantum mechanics gives these selected eigen-values without additional "quantum conditions", as a mathematical consequence of the particular form $H(q,p)$ of the energy function. $H(q,p)$, depending on q and p, represents the energy of a "conservative system", in contrast to the case of a function $H(q,p,t)$ depending on the time t explicitly, which characterizes a non-conservative system.

§ 45. SCHRÖDINGER'S EQUATION FOR NON-CONSERVATIVE SYSTEMS

If we have a set-up in which various values E of the energy are represented by a certain abundance $\sigma(E) = |\chi(E)|^2$, we expect the abundance *amplitude* for finding a value E at the time t to

† See footnote on p. 66.

have the form (see footnote ‡ on p. 67)

$$\phi(t, E) = \chi(E) \,.\, \Psi_E(t) = \chi(E) \,.\, e^{\frac{2\pi i}{h} E.t}.$$

The probability amplitude of finding a particle in the same set-up at the particular point q at the same time t with any energy E whatsoever will then be according to the interference theorem

$$\psi(t, q) = \int \phi(t, E) \,.\, \psi_E(q)\, dE \tag{19}$$

$$= \int \chi(E) \,.\, e^{\frac{2\pi i}{h} E.t} \,.\, \psi_E(q)\, dE,$$

where $\psi_E(q)$ is a standard function, a solution of the Schrödinger equation (18):

$$H(q, \mathbf{p})\, \psi_E(q) = E \,.\, \psi_E(q).$$

If we subject $\psi(t, q)$ of (19) to the operator H, we obtain

$$H(q, \mathbf{p})\, \psi(t, q) = \int \chi(E)\, e^{\frac{2\pi i}{h} E.t} \,.\, H(q, \mathbf{p})\, \psi_E(q)\, dE$$

$$= \int \chi(E)\, e^{\frac{2\pi i}{h} E.t} \,.\, E \,.\, \psi_E(q)\, dE,$$

on account of (18). On the other hand, if we subject $\psi(t, q)$ of (19) to a time differentiation, we obtain

$$\frac{h}{2\pi i} \frac{\partial}{\partial t} \psi(t, q) = \frac{h}{2\pi i} \int \chi(E) \,.\, \frac{2\pi i}{h} E \,.\, e^{\frac{2\pi i}{h} E.t} \,.\, \psi_E(q)\, dE.$$

Since the right-hand sides of the last two equations are equal, the same must be true for their left-hand sides. So we obtain the equation

$$H(q, \mathbf{p})\, \psi(t, q) = \frac{h}{2\pi i} \frac{\partial}{\partial t} \psi(t, q). \tag{20}$$

This fundamental differential equation applies to the density amplitude $\psi(t, q)$ in any conservative set-up, in which various energies are present with constant abundances $\sigma(E) = |\chi(E)|^2$.

Consider now a set-up that is subject to a change in time produced by an external influence, in the form of a variable external force giving rise to a variable potential energy. In this case we have a *non-conservative* energy $H(q, p; t)$ containing t as a parameter, and the abundances $\sigma(E)$ will change in time, too.

We may then suppose that the amplitude $\psi(t, q)$ can still be found, at least approximately, as a solution of the differential equation (20), in which H is now a non-conservative function $H(q, p; t)$:

$$H(q, \mathbf{p}; t)\,\psi(t, q) = \frac{h}{2\pi i}\frac{\partial}{\partial t}\psi(t, q). \tag{21}$$

This last equation is one of the fundamental tools of quantum mechanics for studying the effects of variable external influences on atomic systems.

§ 46. PERTURBATION THEORY

If $H(q, \mathbf{p}; t)$ consists of a dominant conservative term $H^0(q, \mathbf{p})$ plus an additional non-conservative (perturbation) term

$$H(q, \mathbf{p}; t) = H^0(q, \mathbf{p}) + H'(q, \mathbf{p}; t), \tag{22}$$

then we may expand the solution $\psi(t, q)$ of (21) into a series (or into an integral) of the eigen-functions $\psi^0(q)$ of the unperturbed problem

$$H^0(q, \mathbf{p})\,\psi_E^0(q) = E\,.\,\psi_E^0(q)$$

in the form (refer to (19))

$$\psi(t, q) = \int \chi(E, t)\,.\,e^{\frac{2\pi i}{h}E.t}\,.\,\psi_E^0(q)\,dE, \tag{23}$$

where the abundance amplitude $\chi(E, t)$ now depends on the time. The alteration of χ and $\sigma = |\chi|^2$ in time can be interpreted in a corpuscular fashion as being due to transitions of particles from one energy level to another under the influence of the perturbation $H'(q, \mathbf{p}; t)$. This theory will be only approximately correct, since it neglects the reaction of the perturbed system on the perturbing source. In an accurate theory both together form a conservative system subject to the exact equation (20), except that now q in $\psi(t, q)$ represents the co-ordinates of the whole system. For instance, the influence of light on the matter can be treated in an accurate manner only if we regard light and matter together as one complete system. In this way Dirac has developed an exact theory of radiation in a conclusive manner.

§47. ORTHOGONALITY, NORMALIZATION AND HERMITIAN CONJUGACY

We next proceed to derive some properties of the probability amplitudes that are of great importance for the practical applications of quantum theory. For the sake of simplicity we may drop the lower indices K, L, ... as though we had a one-dimensional system. Consider the equation (9):

$$(9')\qquad \beta(q, \mathbf{p})\,\phi_{\beta'}(q) = \beta' . \phi_{\beta'}(q) \quad \left(\mathbf{p} = \frac{h}{2\pi i}\frac{\partial}{\partial q}\right),$$

and also the complex conjugate of (9′) belonging to another value β'' of the function $\beta(q, p)$, which is considered as a real function of its arguments p and q:

$$\beta(q, \mathbf{p}^*)\,\phi^*_{\beta''}(q) = \beta'' . \phi^*_{\beta''}(q) \quad \left(\mathbf{p}^* = -\frac{h}{2\pi i}\frac{\partial}{\partial q}\right).$$

Multiply both sides of the first equation by $\phi^*_{\beta''}(q)$ and both sides of the second equation by $\phi_{\beta'}(q)$, subtract, and integrate over the whole range of q (from $-\infty$ to $+\infty$, or from 0 to 2π if q is an angle, as the case may be):

$$\int [\phi^*_{\beta''} . \beta(q, \mathbf{p})\,\phi_{\beta'} - \phi_{\beta'} . \beta(q, \mathbf{p}^*)\,\phi^*_{\beta''}]\,dq = (\beta' - \beta'') \int \phi_{\beta'}\phi^*_{\beta''}\,dq.$$

If $\beta(q, p)$ is a power series in p like $\sum a_n(q) . p^n$, then $\beta(q, \mathbf{p})$ is a linear differential operator like $\sum a_n(q) . \left(\frac{h}{2\pi i}\right)^n \frac{\partial^n}{\partial q^n}$. We can then transform the volume integral over the range of q on the left into a *surface integral* over the edge of this range. Supposing now that only eigen-functions ϕ are admitted which vanish sufficiently rapidly on the edge, then the integral on the left will be zero. The same holds for the integral on the right if the factor $(\beta' - \beta'')$ is not already zero:

$$(24)\qquad \int \phi_{\beta'}(q) . \phi^*_{\beta''}(q)\,dq = 0 \quad \text{for } \beta' \neq \beta''.$$

This equation expresses the "orthogonality" of any two probability amplitudes $\phi_{\beta'}$ and $\phi_{\beta''}$ belonging to different eigen-values β' and β''.

If $\beta' = \beta''$, then the integral in (24) will be positive, being the integral over $|\phi_{\beta'}(q)|^2$. Since the solution of (9′) is determined only up to a constant factor, we can demand that the integral over $|\phi_{\beta'}(q)|^2$ has the value unity:

$$\int \phi_{\beta'}(q) \,.\, \phi^*_{\beta''}(q)\, dq = 1 \quad \text{for } \beta' = \beta''. \tag{24′}$$

In this case $\phi_{\beta'}(q)$ is said to be "normalized to unity".

Exactly the same considerations can be applied to the eigen-solutions of (13) leading to

$$\int \Phi_{\beta'}(Q) \,.\, \Phi^*_{\beta''}(Q)\, dq = \begin{matrix} 0 & \text{for } \beta' \neq \beta'' \text{ (orthogonality)}, \\ 1 & \text{for } \beta' = \beta'' \text{ (normalization)}. \end{matrix} \tag{25}$$

Furthermore, we may introduce Dirac's δ-function, that is, a probability amplitude $\delta_{\beta'}(\beta)$ which has the value *zero* if the argument β is different from β' and which is normalized to *unity* for $\beta = \beta'$.

According to the interference theorem (5) we then have

$$\int \Phi_{\beta'}(Q) \,.\, X_Q(\beta'')\, dQ = \delta_{\beta'}(\beta'') = \begin{matrix} 0 & \text{for } \beta' \neq \beta'', \\ 1 & \text{for } \beta' = \beta''. \end{matrix} \tag{26}$$

On the other hand, we have from (24) and (24′)

$$\int \Phi_{\beta'}(Q) \,.\, \Phi^*_{\beta''}(Q)\, dQ = \begin{matrix} 0 & \text{for } \beta' \neq \beta'', \\ 1 & \text{for } \beta' = \beta''. \end{matrix} \tag{26′}$$

Comparing the last two equations, we see that if we define

$$X_Q(\beta) = \Phi^*_\beta(Q) \quad \text{(Hermitian conjugacy)} \tag{27}$$

we are in agreement with the general interference theorem (5). Instead of (25), which is an integral over the argument Q, we can now write

$$\int X^*_Q(\beta') \,.\, X_Q(\beta'')\, dQ \begin{matrix} = 0 & \text{for } \beta' \neq \beta'' \text{ (orthogonality)}, \\ = 1 & \text{for } \beta' = \beta'' \text{ (normalization)}, \end{matrix} \tag{28}$$

which is an integral over the lower index.

§ 48. GENERAL MATRIX ELEMENTS

Suppose a certain physical quantity F to be defined as a function $F(q, p)$. We then ask for the average resultant value of F appearing in an optical observation (interpreted by the wave theory of light) on a state which is described in corpuscular terms as a "state of transition" of another physical quantity $\beta(q,p)$ from the values β' to β''. If $\rho_{\beta''\beta'}(F)$ denotes the probability of finding a certain† value F in this transitional state, then we have the average of F in the same state

$$(29) \qquad < F_{\beta''\beta'} > = \int \rho_{\beta''\beta'}(F) \, . \, F \, . \, dF.$$

In analogy to (23) of Part I we obtain the transition density

$$(30) \qquad \rho_{\beta''\beta'}(F) = \chi(\beta'') \, \Phi_{\beta''}(F) \, . \, \chi^*(\beta') \, \Phi^*_{\beta'}(F),$$

where $\chi(\beta')$ and $\chi(\beta'')$ represent the abundance amplitudes of the values β' and β'' present in the special set-up, and where $\Phi_{\beta'}(F)$ and $\Phi_{\beta''}(F)$ are standard amplitudes belonging to pure cases. [The definition (30) must be justified, in the last resort, by observations]. Inserting (30) into (29), we have

$$(31) \qquad < F_{\beta''\beta'} > = \chi(\beta'') \, . \, \chi^*(\beta') \, . \, [F]_{\beta''\beta'},$$

where we have introduced the "matrix element"

$$(32) \qquad [F]_{\beta''\beta'} = \int \Phi^*_{\beta'}(F) \, . \, \Phi_{\beta''}(F) \, . \, F \, . \, dF.$$

The matrix element represents the value of $< F >$ in the standard set-up in which both $\chi(\beta')$ and $\chi(\beta'')$ are *unity*. (32) implies that the matrix elements are "Hermitian" with respect to their indices:

$$(33) \qquad [F]_{\beta''\beta'} = [F]^*_{\beta'\beta''}.$$

In order to calculate $[F]_{\beta''\beta'}$ it would first be necessary to determine the amplitudes $\Phi_{\beta'}(F)$ and $\Phi_{\beta''}(F)$ as solutions of a differential equation like (13) and then combine them to the integral (32). Now if the physical quantities β and F are both defined as functions of certain co-ordinates and momenta q and p, we shall prove the fundamental theorem (Part V) that (32) is identical with the expression

$$(34) \qquad [F]_{\beta''\beta'} = \int \phi^*_{\beta'}(q) \, . \, F(q, \mathbf{p}) \, . \, \phi_{\beta''}(q) \, . \, dq,$$

where $\phi_{\beta'}(q)$ and $\phi_{\beta''}(q)$ are solutions of (9′), and $F(q, \mathbf{p})$ is the operator obtained from the function $F(q,p)$ if p is replaced by

† Refer to footnote † on p. 67.

$\mathbf{p} = \frac{h}{2\pi i}\frac{\partial}{\partial q}$. The physical importance of the matrix elements lies in their invariance with respect to the introduction of new co-ordinates and momenta; that is, one obtains the same matrix elements as in (32), (34) by means of the third formula

$$(35) \qquad [F]_{\beta''\beta'} = \int \phi^*_{\beta'}(Q) . F(Q, \mathbf{P}) . \phi_{\beta''}(Q) . dQ,$$

where $F(Q, \mathbf{P}) = F(q, \mathbf{p})$, is the transformed operator, in terms of any other set of co-ordinates Q and conjugate operators $\mathbf{P}$. The proof is found in Part V.

PART IV

THE PRINCIPLE OF CORRESPONDENCE

§49. CONTACT TRANSFORMATIONS IN CLASSICAL MECHANICS

The term "principle of correspondence" was introduced in Bohr's original theory as a reference to the asymptotic coincidence of spectral frequencies and intensities emitted by real atoms when their electrons jump from one Bohr orbit to another, with the frequencies and intensities emitted according to classical electrodynamics by electrons on those orbits themselves. We should like to use the term "correspondence" in a more general way; one that refers to all analogies and asymptotic coincidences of quantum mechanics with both the classical theory of charged particles of matter and with the classical hydrodynamics of a continuous density serving as a medium for matter waves. Such a correspondence does exist not only with respect to the observed facts but also with respect to the mathematical methods and has proved to be of great heuristic value for the development of quantum mechanics.

As a first example of this correspondence we may consider the manner of passing over from one system of co-ordinates q_K and momenta p_K to another system Q_K and P_K of conjugate variables in classical mechanics, as compared with the transition from one set of co-ordinates q_K and *operators* $\mathbf{p}_K = \frac{h}{2\pi i}\frac{\partial}{\partial q_K}$ to another set Q_K and $\mathbf{P}_K$ in quantum mechanics.

The correspondence of transformations in classical and quantum mechanics has been explained for the special case of point transformations in Part III, §41. In order to generalize the theory we must first give an outline of the *classical* theory of canonical transformations.

A system of mass points (for instance the N electrons of an atom) may be described by $3N$ co-ordinates q_K and momenta p_K

which are called "conjugate" if they satisfy a certain criterion to be described later in equation (4). We may try to describe the same system in terms of $3N$ new co-ordinates Q_L and conjugate momenta P_L. The new momenta P_L shall be certain prescribed functions of the original q and p:

$$P_L = P_L(q, p). \tag{1}$$

On account of (1) the momenta p_K are then certain functions

$$p_K = p_K(q, P). \tag{2}$$

The new Q_L will be calculated as functions of the q and p by the following process called a "contact transformation". First we try to find a "function of action" $S(q, P)$ which satisfies the conditions

$$\frac{\partial S(q, P)}{\partial q_K} = p_K(q, P), \tag{3}$$

whose right-hand sides are identical with those of (2). Then we define the new co-ordinates Q_L by

$$Q_L = \frac{\partial S(q, P)}{\partial P_L}. \tag{3'}$$

Finally we use (3) and (3′) to express the Q_L and P_L in terms of the q and p:

$$Q_L = Q_L(q, p), \quad P_L = P_L(q, p),$$

and conversely

$$q_K = q_K(Q, P), \quad p_K = p_K(Q, P).$$

So far we have applied the letters q, p and Q, P only as mathematical symbols, and have established certain formal relations between them by means of a contact transformation (3), (3′). Now we give them a physical meaning as co-ordinates and momenta by introducing an arbitrary function $H(q, p)$ and supposing that the q's and p's vary in time according to the equations

$$\dot{q}_K = \frac{\partial H}{\partial p_K}, \quad \dot{p}_K = -\frac{\partial H}{\partial q_K}. \tag{4}$$

These are the Hamilton equations of motion of a mechanical system whose *energy* function is $H(q, p)$.

If we now use the transformation $q, p \to Q, P$ and write

$$H(q,p) = H(q(Q,P), p(Q,P)) = \mathscr{H}(Q,P),$$

then it can be proved† that the new co-ordinates and momenta change in time according to the equations

$$\dot{Q}_K = \frac{\partial \mathscr{H}}{\partial P_K}, \quad \dot{P}_K = -\frac{\partial \mathscr{H}}{\partial Q_K}, \tag{4'}$$

which have the same "canonical" form as (4).

† We wish to prove that (4′) is a consequence of (4) because of (3) and (3′). According to (3) and (3′) we have

$$\sum_K (p_K\, dq_K + Q_K dP_K) = \sum_K \left(\frac{\partial S}{\partial q_K} dq_K + \frac{\partial S}{\partial P_K} dP_K \right) = dS. \tag{5}$$

Dividing this by a time increment dt, we have (the dot indicates d/dt)

$$\sum_K (p_K \dot{q}_K + Q_K \dot{P}_K) - \dot{S} = 0.$$

A variation of this equation gives

$$\sum_K (\delta p_K\, \dot{q}_K + p_K\, \delta\dot{q}_K + \delta Q_K \dot{P}_K + Q_K\, \delta\dot{P}_K) - \delta\dot{S} = 0. \tag{5'}$$

Now since

$$p_K\, \delta\dot{q}_K = \frac{d}{dt}(p_K\, \delta q_K) - \dot{p}_K\, \delta q_K,$$

$$Q_K\, \delta\dot{P}_K = \frac{d}{dt}(Q_K\, \delta P_K) - \dot{Q}_K\, \delta P_K,$$

$$\delta\dot{S} = \frac{d}{dt}\delta S,$$

we can write instead of (5′)

$$\sum_K \frac{d}{dt}(p_K\, \delta q_K + Q_K\, \delta P_K - \delta S) + (-\dot{p}_K\, \delta q_K + \dot{q}_K\, \delta p_K - \dot{Q}_K\, \delta P_K + \dot{P}_K\, \delta Q_K) = 0.$$

The first bracket vanishes because of (5), so we are left with

$$\sum_K (-\dot{p}_K\, \delta q_K + \dot{q}_K\, \delta p_K) = \sum_K (-\dot{P}_K\, \delta Q_K + \dot{Q}_K\, \delta P_K).$$

Using (4), the left-hand side reduces to δH. If furthermore $H(q,p) = \mathscr{H}(Q,P)$, then we have

$$\delta H = \delta\mathscr{H} = \sum_K \left(\frac{\partial \mathscr{H}}{\partial Q_K} \delta Q_K + \frac{\partial \mathscr{H}}{\partial P_K} \delta P_K \right).$$

Since the left-hand sides of the last two equations are identical, the same applies to the right-hand sides for any variations δQ_K and δP_K. So we are led to the equations

$$-\dot{P}_K = \frac{\partial \mathscr{H}}{\partial Q_K} \quad \text{and} \quad \dot{Q}_K = \frac{\partial \mathscr{H}}{\partial P_K},$$

identical with (4′) as a consequence of (3), (3′). Thus Hamilton's equations of motion are invariant with respect to a contact transformation (3) and (3′). All these considerations will hold if we replace the letters q and p by Q and P and vice versa.

§50. POINT TRANSFORMATIONS

As the simplest example of a contact transformation we now discuss point transformations, where the Q_K are functions of the q's only, not containing the p's. In this case we may write the function of action $S(q, P)$ immediately in the form

$$S(q, P) = \sum_L P_L \, . \, Q_L(q), \tag{6}$$

which indeed satisfies (3′):

$$Q_L = \frac{\partial S}{\partial P_L}. \tag{6′}$$

From (3), (6) we then obtain

$$p_K = \frac{\partial S}{\partial q_K} = \sum_L \frac{\partial S}{\partial Q_L} \cdot \frac{\partial Q_L}{\partial q_K} = \sum_L \frac{\partial Q_L}{\partial q_K} \, . \, P_L. \tag{6″}$$

We notice the close correspondence of these equations with those defining a point transformation of co-ordinates q and *operators* $\mathbf{p}$ into new Q and $\mathbf{P}$ as introduced in Part III, §41, the only difference being that the momenta p_K and P_K are replaced by the operators

$$\mathbf{p}_K = \frac{h}{2\pi i} \frac{\partial}{\partial q_K} \quad \text{and} \quad \mathbf{P}_K = \frac{h}{2\pi i} \frac{\partial}{\partial Q_K}. \tag{7}$$

§51. CONTACT TRANSFORMATIONS IN QUANTUM MECHANICS

This correspondence leads to a general definition of contact transformations $q, \mathbf{p} \rightarrow Q, \mathbf{P}$ in quantum mechanics. If the new co-ordinates Q_L are prescribed in the form $Q_L(q, \mathbf{p})$ and the old co-ordinates are obtained conversely in the form $q_K = q_K(Q, \mathbf{p})$, we can find new "conjugate" operators $P_L = P_L(q, \mathbf{p})$ by the following process. First we must find an "operator of action" $S(Q, \mathbf{p})$ such that the equations

$$\frac{\partial S(Q, \mathbf{p})}{\partial \mathbf{p}_K} = q_K \tag{8}$$

define the new co-ordinates Q_L in the prescribed form of $Q_L(q, \mathbf{p})$. By differentiation with respect to the operator $\mathbf{p}_K$ is meant a formal differentiation (for instance $\frac{\partial}{\partial \mathbf{p}_K}\mathbf{p}_K^2$ shall stand for $2\mathbf{p}_K$, as if $\mathbf{p}_K$ were an ordinary quantity). Now we define the new conjugate operators $\mathbf{P}_L$ by the equations

$$\frac{\partial S(Q,\mathbf{p})}{\partial Q_L}=\mathbf{P}_L \tag{8'}$$

as functions of the Q's and the operators $\mathbf{p}$. Finally we use (8) and (8′) for expressing Q and $\mathbf{P}$ in terms of q and $\mathbf{p}$, or vice versa q and $\mathbf{p}$ in terms of the Q and $\mathbf{P}$.

In order to illustrate this procedure we use once more the case of a *point transformation*, where q_K is prescribed in the form of $q_K(Q)$. Here we must take

$$S(Q,\mathbf{p})=\sum_K q_K(Q)\cdot\mathbf{p}_K, \tag{9}$$

so that S indeed satisfies (8):

$$\frac{\partial S}{\partial \mathbf{p}_K}=q_K(Q).$$

The operator $\mathbf{P}_L$ is then obtained from (8′) in the form of

$$\mathbf{P}_L=\frac{\partial S}{\partial Q_L}=\sum_K \frac{\partial q_K(Q)}{\partial Q_L}\cdot\mathbf{p}_K, \tag{9'}$$

as a function of the q and $\mathbf{p}$.

The significance of the contact transformations q, $\mathbf{p}\rightarrow Q$, $\mathbf{P}$ defined in (8) and (8′) lies in the following fact. If an operator $\beta(q,\mathbf{p})$ is transformed into the form

$$\beta(q,\mathbf{p})=\beta(q(Q,\mathbf{P}),p(Q,\mathbf{P}))=\mathrm{B}(Q,\mathbf{P})$$

by means of a contact transformation (8) and (8′), then we may prove (§ 64, Part V) that the eigen-functions of $\beta(q,\mathbf{p})$ and $\mathrm{B}(Q,\mathbf{P})$ satisfy the fundamental *interference theorem* A of § 42. Furthermore, we are going to prove in Part V, § 65 the invariance of the matrix elements of any physical function $F(q,p)$ with respect to *contact transformations*; so the matrix elements of F prove to have a meaning independent of the special set of co-ordinates and conjugate momenta (q, p or Q, P) used to define the physical function F.

§52. CONSTANTS OF MOTION AND ANGULAR CO-ORDINATES

Another important example of a contact transformation in classical mechanics is the transition from original q and p to such new co-ordinates α and conjugate momenta β that the β_K are "constants of the motion" which do not change in time. This is the case if the α and β are such functions of the q and p that the energy function $H(q,p)=\mathscr{H}(\alpha,\beta)$ becomes a function of the β alone:

$$H(q,p)=\mathscr{H}(\beta).$$

Indeed (4′) then leads to

$$\text{(10)} \qquad \dot{\beta}_K=-\frac{\partial\mathscr{H}}{\partial\alpha_K}=0, \quad \text{hence} \quad \beta_K=\text{const.};$$

at the same time (4′) gives

$$\text{(10′)} \quad \dot{\alpha}_K=\frac{\partial\mathscr{H}}{\partial\beta_K}=\text{const.}=\nu_K, \quad \text{hence} \quad \alpha_K=\nu_K t+\alpha_K^0,$$

that is, the α_K increase linearly with the time like the angle of a uniform rotation; the α_K are called *angular variables* conjugate to the *constants of the motion* β_K. For given values β_K' of the constants of the motion the function of action $S(q,\beta')$ will have various values

$$\text{(11)} \qquad S(q,\beta')=S$$

at various points q of the space. Conversely, a fixed value S on the right side of the last equation is realized in a manifold of points q representing a *surface* in space. Fixing once for all the values β_K', and giving S on the right side of (11) various current values S_1, S_2, ..., we obtain a succession of surfaces $S(q,\beta')=$ const. in space (Fig. 13).

$S=S_1$ S_2 ··

Fig. 13.

The orthogonal trajectories to this succession of surfaces represent mechanical orbits belonging to the fixed set of constants of motion β_K'. Indeed the direction in space orthogonal to the surface (11) has the components

$$\frac{\partial S(q,\beta')}{\partial q_K} \quad (K=1,2,\dots),$$

which are identical with the components p_K of the momentum vector according to (3) and are thus parallel to the velocity vector of a mechanical orbit. This orthogonality of the orbits to the surfaces $S(\alpha,\beta')=S$ implies that the line integral

$$(12) \qquad \int dS=\int \sum_K \frac{\partial S(q,\beta')}{\partial q_K} dq_K=\int \sum_K p_K \cdot dq_K$$

between two points A and B of the same mechanical orbit is a *minimum* if taken along the orbit itself compared with the integral along any other (dotted) line between A and B (Fig. 14). This is the *principle of least action*. Its significance lies in the fact that the same minimum value of the integral (12) along the same orbit between the same two points can be expressed in terms of any other canonical variables Q and P, for instance in terms of the angular variables α and β†

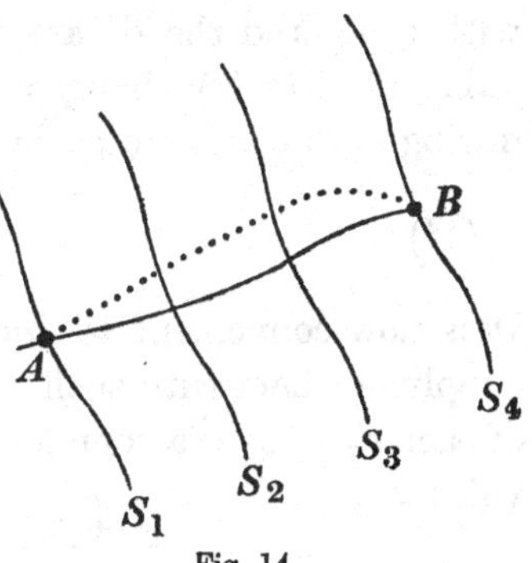

Fig. 14.

$$(13) \qquad \sum_K \int p_K \cdot dq_K = \sum_K \beta_K \int d\alpha_K = \sum_K \int P_K \cdot dQ_K = \int dS.$$

† In order to prove the first part of equation (13) remember that according to (3) and (3′) for $S(q,\beta)$

$$\frac{\partial S}{\partial q_L}=p_L, \quad \frac{\partial S}{\partial \beta_K}=\alpha_K, \quad \text{hence} \quad \frac{\partial \alpha_K}{\partial q_L}=\frac{\partial^2 S}{\partial \beta_K \partial q_L}=\frac{\partial p_L}{\partial \beta_K}.$$

Multiplying by dq_L summing over all values L, and integrating, gives

$$\int d\alpha_K=\int \sum_L \frac{\partial \alpha_K}{\partial q_L} dq_L=\int \sum_L \frac{\partial p_L}{\partial \beta_K} dq_L=\frac{\partial}{\partial \beta_K}\int \sum_L p_L \cdot dq_L.$$

This is a differential equation for $\int \sum_L p_L \cdot dq_L$, and its solution is

$$\int \sum_L p_L \cdot dq_L = \sum_K \beta_K \int d\alpha_K.$$

The second part of (13) is then proved in exactly the same way.

§ 53. PERIODIC ORBITS

Of particular interest to atomic physics and quantum theory are periodic orbits. They gave the first evidence of a quantization in the original theory of Planck, Bohr and Sommerfeld, and their counterpart is found in the quantized states of quantum mechanics.

Suppose the values of the space co-ordinates q_K to repeat after certain commensurable time intervals. Then the q_K will increase part of the time, in other parts of the period they will decrease, whereas the angular co-ordinates $\alpha_K = \nu_K t + \alpha_K^0$ increase uniformly with time, and the β_K are constant in time. Furthermore, the value of $S(q, \beta')$ changes uniformly along an orbit, so that during each complete cycle S increases by the same amount:

$$\oint dS = \sum_K \beta_K \oint d\alpha_K. \tag{14}$$

It is now convenient to *normalize* the angular variables α_K by supplying them with such constant factors c_K that the integrals of each $c_K \alpha_K$ over a complete cycle of the whole system have the value 2π:

$$c_K \oint d\alpha_K = 2\pi \text{ for each } K.$$

Introducing the new angular variables γ_K,

$$\gamma_K = c_K \alpha_K = c_K(\nu_K t + \alpha_K^0) \tag{15}$$

gives

$$\oint d\gamma_K = 2\pi.$$

At the same time one introduces new constants of the motion

$$J_K = \beta_K / c_K \tag{15'}$$

so that

$$\beta_K \oint d\alpha_K = J_K \oint d\gamma_K.$$

With these *normal variables*, we have

$$\oint dS = \sum_K \beta_K \oint d\alpha_K = \sum_K J_K \oint d\gamma_K = \sum_K 2\pi J_K \tag{16}$$

as the action integral over a whole periodic orbit.

The quantum theory of Planck, Bohr and Sommerfeld was based on the assumption that only such periodic orbits really occur as are characterized by quantum conditions for the normalized constants of the motion J_K, namely

$$J'_K = n'_K \frac{h}{2\pi}, \tag{17}$$

where the n'_K are integers. The function of action S then increases according to (16) by the amount

$$\oint dS = \sum n'_K h = n'h \tag{17'}$$

along every closed mechanical orbit characterized by the set (17) of quantized constants of the motion. Conversely, the value of $S(q, \beta')$ at a certain point q of a periodic orbit is determined only up to an additive constant $n'h$, if the constants of the motion J_K are quantized according to (17).

§ 54. DE BROGLIE AND SCHRÖDINGER FUNCTION; CORRESPONDENCE TO CLASSICAL MECHANICS

A bridge from these results of classical mechanics to quantum mechanics has been erected by L. de Broglie. He took the function of action $S(q, \beta')$ of a quantized system of mechanical orbits as the exponent of a complex exponential function:

$$\psi_{\beta'}(q) = e^{\frac{2\pi i}{h} S(q, \beta')}. \tag{18}$$

Passing along a periodic orbit the function S increases by $n'h$ according to (17'); hence the exponent of de Broglie's function (18) increases by $2\pi i n'$ and thus $\psi_{\beta'}(q)$ itself repeats its initial value n' times during the cycle. Although the value of the function $S(q, \beta')$ is determined at every point q only up to $n'h$, the function ψ is a *unique function* of the space co-ordinates. According to de Broglie the rules (17) of quantization state: Only such values of the constants of motion J'_K or β'_K are to be admitted as render the function $\psi_{\beta'}(q)$ in (18) a *unique* function in space.

The de Broglie wave function (18) with the classical function of action $S(q, \beta')$ as exponent is approximately, but not quite, identical with the Schrödinger function $\psi_{\beta'}(q)$ as required by quantum mechanics. In the latter theory $\psi_{\beta'}(q)$ is supposed to be a solution of the simultaneous differential equations

$$(19) \qquad \beta_K(q, \mathbf{p})\,\psi_{\beta'}(q) = \beta'_K \cdot \psi_{\beta'}(q), \qquad (K = 1, 2, \ldots)$$

where the $\beta_K(q, \mathbf{p})$ are the constants of the motion and β'_K fixed values of them. But the solutions of (19) are approximately equal to (18). The correspondence may be demonstrated in the particular case that β_L is the *energy* function of a mass point m in the potential field $U(q)$:

$$(19') \qquad \beta_L(q, p) = \frac{1}{2m} p^2 + U(q) = E,$$

hence
$$\beta_L(q, \mathbf{p}) = \frac{1}{2m}\left(\frac{h}{2\pi i}\right)^2 \frac{\partial^2}{\partial q^2} + U(q),$$

so that the Lth of the equations (19) reads

$$(19'') \qquad \frac{1}{2m}\left(\frac{h}{2\pi i}\right)^2 \frac{\partial^2}{\partial q^2}\psi + U(q)\,\psi = E\psi,$$

with $\beta'_L = E$ as a constant value of the energy. Now if we insert the de Broglie function (18) for ψ into (19'') and carry out the differentiations, we obtain

$$\frac{1}{2m}\left\{\left(\frac{\partial S}{\partial q}\right)^2 + \frac{h}{2\pi i}\frac{\partial^2 S}{\partial q^2}\right\}\psi + U(q)\,\psi = E\psi.$$

Dividing by the factor $\psi/2m$ and replacing $2m(E - U)$ by p^2 according to (19'), we obtain the result: If ψ is assumed to have the exponential form $\psi = e^{\frac{2\pi i}{h} S}$, then the function S would have to satisfy the equation

$$(20) \qquad \left(\frac{\partial S}{\partial q}\right)^2 = p^2 - \frac{h}{2\pi i}\frac{\partial^2 S}{\partial q^2},$$

where $p^2 = 2m\{E - U(q)\}$. The function S of classical mechanics, however, would have to satisfy the equation (3) or

$$(20') \qquad \left(\frac{\partial S}{\partial q}\right)^2 = p^2.$$

The difference between the last two equations is only of the order of h. de Broglie's function (18), using the classical S of (20′) with quantized values (17) of the constants of motion, proves to be an approximation to the new quantum theory. Kramers, Wentzel and Brillouin have utilized this feature of the de Broglie function in order to get a method for solving the equation (19) of quantum mechanics by successive approximations.

§ 55. PACKETS OF PROBABILITY

A narrow bundle of mechanical orbits (Fig. 15), not necessarily periodic but all of them belonging to the same constants β' of the motion, may emanate from a small surface element Δs_0 of the surface $S(q, \beta') = S_0$. According to the rules of *classical* mechanics the orbits will pierce subsequent surfaces $S(q, \beta') = S_1, S_2, \ldots$ perpendicularly with well-defined cross-sections $\Delta s_1, \Delta s_2, \ldots$. We can then define a de Broglie function

$$(21)\qquad \begin{cases} \psi_{\beta'}(q) = e^{\frac{2\pi i}{h} S(q, \beta')} & \text{within the } \Delta s, \\ \psi_{\beta'}(q) = 0 & \text{outside of the } \Delta s, \end{cases}$$

representing a well-defined bundle of "rays" limited to the respective cross-sections $\Delta s_0, \Delta s_1, \Delta s_2, \ldots$, like a bundle of light rays in geometrical optics.

The classical wave ray (21) will, however, not be a solution of the differential equation (19). On the contrary, this equation will have a solution $\psi_{\beta'}(q)$ that extends through the whole of space with an infinite cross-section. The question arises, however, whether it is possible to *superpose* the solutions $\psi_{\beta'}(q)$ of many differential equations (19), each of them belonging to slightly different constants β', in order to obtain a "packet" of solutions

$$(22)\qquad \sum_{\beta'} A_{\beta'} \psi_{\beta'}(q) = \psi(q),$$

so that the function $\psi(q)$ vanishes except for a finite cross-section Δs_0 (Fig. 15) at least on the surface S_0. This is indeed possible if the factors $A_{\beta'}$ in the sum or integral (22) are given suitable values. And quite in line with the more special considerations that led to the uncertainty relations for p and q in Part III, it turns out that the smaller we want the original cross-section Δs_0 of the packet (22) to be, the wider must be the range $\Delta\beta'$ for the values of the constants of motion to which we have to resort in the sum (22); in other words, the more heterogeneous will be the packet. As a consequence of this heterogeneity the cross-section of the packet (22) will not remain as small as Δs_1, Δs_2, ... in its subsequent course. Instead it will be *diffracted* more and more (dotted lines of Fig. 15) to wider cross-sections, quite in contrast with the classical ray (21) which keeps within the boundaries of the "geometrical" shadow of the original opening Δs_0.

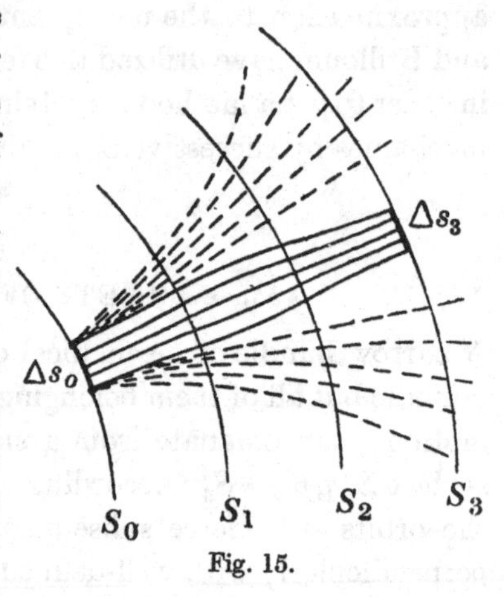

Fig. 15.

§ 56. CORRESPONDENCE TO HYDRODYNAMICS

If we interpret the ψ-function as the amplitude of a continuous density $\rho = |\psi|^2$, then it turns out that the rules of hydrodynamics apply only if we assume certain "non-mechanical" forces to be present in the fluid (E. Madelung (19)). Let us consider the case of a set-up in which various values of the energy of single particles are present with various abundances (mixed case, packet of pure solutions) so that ψ as a function of the space coordinates q and the time is a solution of the equation (21), § 45:

$$(23) \qquad H(q, \mathbf{p}, t)\,\psi = -\frac{h}{2\pi i}\frac{\partial}{\partial t}\psi.$$

In particular, let the energy function H have the form

$$H(q,p,t) = \frac{1}{2m}(p_x^2 + p_y^2 + p_z^2) + U(x,y,z,t),$$

where the potential energy U may be an explicit function of the time, representing a non-conservative (external) influence in addition to conservative (internal) forces. $\psi(xyzt)$ must then satisfy the equation (as a special case of (23))

$$\Delta\psi - \frac{8\pi^2 m}{h^2} U \,.\, \psi = -\frac{4\pi i m}{h} \frac{\partial\psi}{\partial t}. \tag{24}$$

At the same time the complex conjugate function ψ^* satisfies the equation

$$\Delta\psi^* - \frac{8\pi^2 m}{h^2} U \,.\, \psi^* = +\frac{4\pi i m}{h} \frac{\partial\psi^*}{\partial t}. \tag{24'}$$

Multiplying (24) by ψ^* and (24′) by ψ and subtracting, we obtain† the equation

$$\frac{h}{4\pi i} \operatorname{div}(\psi^* \operatorname{grad}\psi - \psi \operatorname{grad}\psi^*) + \frac{\partial}{\partial t}(m\psi\psi^*) = 0. \tag{25}$$

Now in hydrodynamics we have the *equation of continuity*

$$\operatorname{div} j + \frac{\partial\rho}{\partial t} = 0 \qquad \begin{array}{l} \rho = \text{mass density,} \\ j = \text{vector of the current density.} \end{array}$$

This suggests that we interpret

(26) $m\psi\psi^* = m\,|\,\psi\,|^2 = \rho$ as mass density,

(26′) $\frac{h}{4\pi i}(\psi^* \operatorname{grad}\psi - \psi \operatorname{grad}\psi^*) = j$ as vector of the current density,

of the fluid in space and time, supposing that ψ is normalized so that $\int |\,\psi\,|^2 dv = 1$.

The continuity relation, integrated over all space, gives

$$\frac{\partial}{\partial t}\int \rho\, dv + \int \operatorname{div} j \,.\, dv = 0.$$

The second integral is equivalent to a surface integral $\int j_n\, ds$ over the infinitely distant surface and gives the result zero, if j is con-

† Using the vector formulae

$$\Delta u = \operatorname{div}(\operatorname{grad} u),$$
$$u \,.\, \Delta v = \operatorname{div}(u \,.\, \operatorname{grad} v) - (\operatorname{grad} u) \,.\, (\operatorname{grad} v).$$

fined to finite values. So we have the *conservation of the total mass* in time:

$$(27) \qquad \frac{\partial}{\partial t}\int \rho\, dv = 0, \quad \text{that is} \quad \frac{\partial}{\partial t}\int m\,|\psi(q,t)|^2\, dv = 0.$$

If we define the current velocity c in the fluid by

$$(28) \qquad c = \frac{j}{\rho} = \frac{h}{4\pi i m}\left(\frac{\operatorname{grad}\psi}{\psi} - \frac{\operatorname{grad}\psi^*}{\psi^*}\right),$$

and use the differential equations (24) and (24′) for ψ and ψ^*, we obtain with E. Madelung the following equation:

$$-\operatorname{grad} U + \frac{h^2}{8\pi^2 m}\operatorname{grad}\frac{\Delta\sqrt{\psi\psi^*}}{\sqrt{\psi\psi^*}} - \frac{m}{2}\operatorname{grad} c^2 = m\frac{\partial c}{\partial t}.$$

Introducing instead of the partial differential quotient $\partial c/\partial t$ the total differential quotient

$$\frac{dc}{dt} = \frac{\partial c}{\partial t} + \frac{1}{2}\operatorname{grad} c^2 = \frac{\partial c}{\partial t} + (c \operatorname{grad})\, c,$$

we obtain the equation of motion

$$(29) \qquad -\operatorname{grad}\left(U - \frac{h^2}{8\pi^2 m}\frac{\Delta\sqrt{\rho}}{\sqrt{\rho}}\right) = m\frac{dc}{dt}.$$

The forces that accelerate the velocity c are due first to the potential U and second to the additional "internal" potential

$$(29') \qquad -\frac{h^2}{8\pi^2 m}\frac{\Delta\sqrt{\rho}}{\sqrt{\rho}} = U^i$$

that must be introduced in order to explain the behaviour of the fluid in a mechanical way. Such an additional potential is necessary if we wish to explain in a mechanical manner that ψ has a finite value in ranges where the ordinary potential energy $U(q)$ subtracted from the given total energy E would lead to a negative kinetic energy $\frac{1}{2}mc^2$ and to an imaginary velocity c were it not for the additional internal potential, which makes c real at every point.

§57. MOTION AND SCATTERING OF WAVE PACKETS

If the density $\rho = |\psi(q,t)|^2$ is condensed, at the time $t=0$, in a small range around a point P_0, we speak of a "wave packet" in so far as the density amplitude $\psi(q,0)$ can be built up as a superposition of many functions $\psi_E(q)$ belonging to slightly different energy values E. It is interesting to follow the density distribution of such a density maximum for later times $t>0$ according to the differential equation (24). The result is a gradual flattening of the density maximum to a wider and wider range. The rate of this process depends however on the half-width of the original maximum. If, for instance, ρ is condensed at $t=0$ along a diameter of 10^{-8} cm. and m is supposed to be about 10^{-24} gr. (H-atom), then the diameter of the maximum will increase to twice its size after about 10^{-13} sec. (as a result of (24)). If on the other hand the initial diameter is 0·1 cm. and the mass 10^{-3} gr., then the diameter will reach twice its size only after 10^{15} years.

One can understand this flattening process of a narrow density maximum of the wave function ψ from the corpuscular point of view of the uncertainty principle. The narrow packet in space means an all-the-wider range of uncertainty of corresponding momenta and hence an all-the-wider range of velocities, leading to a spread of the matter into all directions.

P. Ehrenfest (20) has found the interesting result that the centre of gravity of such a density maximum moves like a mass particle according to the rules of classical mechanics. Indeed, if we multiply (29) by ρ and integrate over all space, we obtain

$$\int \frac{d}{dt}(mc)\,.\,\rho\, dv = \int (-\operatorname{grad} U)\,\rho\, dv + \int (-\operatorname{grad} U^i)\,\rho\, dv.$$

The last integral over the non-mechanical internal force (cf. (29′)) can be transformed, however, into a surface integral and vanishes if ρ decreases sufficiently at infinity, so that we are left with

$$\int \frac{d}{dt}(mc)\,.\,\rho\, dv = \int (-\operatorname{grad} U)\,.\,\rho\, dv. \tag{30}$$

That means, however, that the centre of gravity of the wave packet moves only under the influence of the external force $(-\operatorname{grad} U)$.

§58. FORMAL CORRESPONDENCE BETWEEN CLASSICAL AND QUANTUM MECHANICS

The correspondence between classical and quantum mechanics may be finally expressed by the following comparison:

Classical Mechanics

For fixed constants of motion β'_K we obtain a bundle of mechanical orbits with different starting points in space but all of them controlled by the Hamilton equations of motion:

$$\dot{q}_K = \frac{\partial H}{\partial p_K}, \quad \dot{p}_K = -\frac{\partial H}{\partial q_K}.$$

If new co-ordinates $Q_L(q,p)$ are introduced, then we may describe the same set of orbits in terms of the transformed co-ordinates Q, using the direct formula of transformation

$$Q_L(t) = Q_L(q(t), p(t))$$

in terms of the original $q_K(t)$ and $p_K(t)$.

But we can obtain the same set of orbits in terms of the new variables directly as solutions of Hamilton's equations

$$\dot{Q}_K = \frac{\partial \mathscr{H}}{\partial P_K}, \quad \dot{P}_K = -\frac{\partial \mathscr{H}}{\partial Q_K}$$

if $\mathscr{H}(Q,P)$ is the transformation of $H(q,p)$, supposing that the transformation $q, p \rightarrow Q, P$ was a contact transformation.

Quantum Mechanics

For fixed values β'_K of physical functions $\beta_K(q,p)$ defining a "pure case" we obtain a certain density amplitude $\phi_{\beta'}(q)$ in various points of space controlled by the differential equation

$$\beta_K(q, \mathbf{p})\, \phi_{\beta'}(q) = \beta'_K \cdot \phi_{\beta'}(q).$$

If new co-ordinates $Q_L(q,p)$ are introduced, we obtain the new density amplitude $\Phi_{\beta'}(Q)$ for the same pure case as the interference integral

$$\Phi_{\beta'}(Q) = \int \phi_{\beta'}(q) \cdot \Psi^*_Q(q)\, dq$$

in terms of the original amplitude $\phi_{\beta'}(q)$.

But we can obtain the same amplitude $\Phi_{\beta'}(Q)$ directly as eigen-function of the differential equations

$$\mathrm{B}_K(Q, \mathbf{P})\, \Phi_{\beta'}(Q) = \beta'_K \cdot \Phi_{\beta'}(Q)$$

if $\mathrm{B}_K(Q, \mathbf{P})$ is the transformation of $\beta_K(q, \mathbf{p})$, supposing that the transition $q, \mathbf{p} \rightarrow Q, \mathbf{P}$ was a contact transformation of quantum mechanics (as developed in Part V).

A correspondence between classical and quantum mechanics is found only in case the functions $\beta_K(q,p)$ are "constants of the motion", although the *calculus* of quantum mechanics may be applied to any physical (real) function $\beta(q,p)$ whatsoever. In practice, however, one can make quantitative observations only under circumstances which are characterized by constant values or by transitions between constant values of such functions $\beta(q,p)$ as have the property of being constants of the motion.

PART V

MATHEMATICAL APPENDIX: PRINCIPLE OF INVARIANCE

59. THE GENERAL THEOREM OF TRANSFORMATION

The main problem of quantum mechanics consists in predicting probability amplitudes like $\phi_{\beta'}(q)$ or $\psi_{q'}(Q)$ or $\Phi_{\beta'}(Q)$. Here $\beta(q,p)$ and $Q(q,p)$ are physical (real) quantities expressed as functions of certain original co-ordinates q and momenta p, and q', β' stand for fixed values of q and β, and $\Phi_{\beta'}(Q)$ means the amplitude of the probability of finding a value Q of $Q(q,p)$ in a set-up in which $\beta(q,p)$ has the fixed value β'. The principal result of quantum physics is the theorem that the various probability amplitudes are in mutual interdependence satisfying equations like

$$\phi_{\beta'}(q) = \int \Phi_{\beta'}(Q) \,.\, \psi_Q(q)\, dQ, \tag{1}$$

supposing that the correct probability amplitudes are used. The latter are to be calculated as eigen-functions of the following differential equations:

$$\beta(q, \mathbf{p})\,\phi_{\beta'}(q) = \beta' \,.\, \phi_{\beta'}(q), \tag{2}$$

$$Q(q, \mathbf{p})\,\psi_{Q'}(q) = Q' \,.\, \psi_{Q'}(q), \tag{3}$$

$$\mathrm{B}(Q, \mathbf{P})\,\Phi_{\beta'}(Q) = \beta' \,.\, \Phi_{\beta'}(Q). \tag{4}$$

Here $\mathbf{p}_K$ stands for $\dfrac{h}{2\pi i}\dfrac{\partial}{\partial q_K}$, and $\mathbf{P}_K$ for $\dfrac{h}{2\pi i}\dfrac{\partial}{\partial Q_K}$, and the operator $\mathrm{B}(Q, \mathbf{P})$ is the transformed operator $\beta(q, \mathbf{p})$:

$$\beta(q, \mathbf{p}) = \beta(q(Q, \mathbf{P}), \mathbf{p}(Q, \mathbf{P})) = \mathrm{B}(Q, \mathbf{P}), \tag{5}$$

by virtue of a "contact transformation" $q, \mathbf{p} \rightarrow Q, \mathbf{P}$ (§ 51 of Part IV) with the help of an operator of action $S(Q, \mathbf{p})$ chosen so that the $q_K(Q, \mathbf{p})$ and $\mathbf{p}_K(Q, \mathbf{P})$ satisfy identically the equations

$$\frac{\partial S(Q, \mathbf{p})}{\partial \mathbf{p}_K} = q_K, \quad \frac{\partial S(Q, \mathbf{p})}{\partial Q_K} = \mathbf{P}_K. \tag{6}$$

We must still give the mathematical proof of the theorem A: *The eigen-functions of* (2), (3), *and* (4) *satisfy the interference rule* (1) *by virtue of the contact transformation* (6).

This theorem is of great generality because the special form of the new variables Q as functions of q and $\mathbf{p}$ does not matter at all. If for example $M(q, p)$ is any other arbitrary physical quantity, $M(q, \mathbf{p})$ the corresponding operator, and $N = N(q, \mathbf{p})$ the conjugate operator obtained by a contact transformation, then we have the interference rule

$$(1') \qquad \phi_{\beta'}(q) = \int \Phi_{\beta'}(M) \,.\, \psi_M(q)\, dM,$$

where $\Phi_{\beta'}(M)$ and $\psi_M(q)$ are eigen-functions of the differential equations

$$(3') \qquad M(q, \mathbf{p})\, \psi_{M'}(q) = M' \,.\, \psi_{M'}(q),$$

$$(4') \qquad \mathrm{B}(M, \mathrm{N})\, \Phi_{\beta'}(M) = \beta' \,.\, \Phi_{\beta'}(M).$$

This independence of the interference theorem from the special set of conjugate variables Q, P or M, N, its *invariance* with respect to transitions from one to another set of conjugate variables, is the reason for giving Part V the title, the *principle of invariance*. In addition, we shall prove the *invariance* of the matrix elements of any physical function F with respect to the special choice of the conjugate variables used.

The following developments are mainly mathematical. In particular we introduce the Born-Jordan-Dirac *calculus of operators*, in order to prove theorem A. We then show with P. Jordan that the contact transformation of (6) by means of an "operator of action" $S(Q, \mathbf{p})$ is equivalent to another transformation made by means of a "transformer function" $T(q, \mathbf{p})$; the latter is easier to deal with, since it contains only the original q and $\mathbf{p}$, instead of being a "mixed" function $S(Q, \mathbf{p})$ of the old operators $\mathbf{p}$ and the new co-ordinates Q. With the help of the transformer function T effecting the contact transformation $\mathbf{p}, q, Q, \mathbf{P}$, one can prove the theorem A as will be shown in § 64.

§60. OPERATOR CALCULUS

Let us derive the simplest rules that apply to calculations in which linear differential operators like

$$\mathbf{p}_K = \frac{h}{2\pi i}\frac{\partial}{\partial q_K}$$

occur. Since by operating on a function of the q we have

$$\frac{\partial}{\partial q_K} + \frac{\partial}{\partial q_L} = \frac{\partial}{\partial q_L} + \frac{\partial}{\partial q_K},$$

we may write

$$\mathbf{p}_K + \mathbf{p}_L = \mathbf{p}_L + \mathbf{p}_K,$$

which represents the rule of commutative addition for linear differential operators. In the same way we have the rule of association and dissociation:

$$(\mathbf{p}_K + \mathbf{p}_L) + \mathbf{p}_M = \mathbf{p}_K + (\mathbf{p}_L + \mathbf{p}_M).$$

Operating on a function $f(q)$ we have

$$\frac{\partial}{\partial q_K}\left(\frac{\partial}{\partial q_L} + \frac{\partial}{\partial q_M}\right) = \frac{\partial^2}{\partial q_K \partial q_L} + \frac{\partial^2}{\partial q_K \partial q_M},$$

which can be written in the abbreviated form of a "product rule"

$$\mathbf{p}_K(\mathbf{p}_L + \mathbf{p}_M) = \mathbf{p}_K\mathbf{p}_L + \mathbf{p}_K\mathbf{p}_M.$$

Also, operating on a function $f(q)$, we have

$$\frac{\partial}{\partial q_K}\left(\frac{\partial^2}{\partial q_L \partial q_M}\right) = \left(\frac{\partial^2}{\partial q_K \partial q_L}\right)\frac{\partial}{\partial q_M}$$

or

$$\mathbf{p}_K(\mathbf{p}_L\mathbf{p}_M) = (\mathbf{p}_K\mathbf{p}_L)\mathbf{p}_M.$$

Here we have the rules of association and dissociation for "products" of operators. It is not possible, however, to apply the rule of *commutation* to "products" of operators. Since $\frac{\partial q_K}{\partial q_K} = 1$ we obtain by partial differentiation

$$\frac{h}{2\pi i}\frac{\partial}{\partial q_K} q_K \cdot f(q) = \frac{h}{2\pi i} f(q) + q_K \frac{h}{2\pi i}\frac{\partial}{\partial q_K} f(q), \tag{7}$$

so we must write the operator equation

$$\mathbf{p}_K \cdot q_K = q_K \cdot \mathbf{p}_K + \frac{h}{2\pi i}. \tag{8}$$

That is, $\mathbf{p}_K q_K$ is different from $q_K \mathbf{p}_K$. On the other hand, since $\frac{\partial q_K}{\partial q_L}=0$, we have

$$\mathbf{p}_K q_L = q_L \mathbf{p}_K \quad \text{for } K \neq L. \tag{8'}$$

According to (8) and (8′) $\mathbf{p}_K$ is commutative with q_L but not with its own conjugate q_K. We see that we can make calculations with operators in the same way as if they were ordinary algebraic quantities subject to the rules of association and dissociation as well as commutation of sums, without commutation of products, since $\mathbf{p}_K q_K$ is different from $q_K \mathbf{p}_K$. Instead, we must use the "exchange relation" (8) of Heisenberg.

§61. EXCHANGE RELATIONS; THREE CRITERIA FOR CONJUGACY

In the same way we may write the operator equation

$$\mathbf{P}_K Q_K = Q_K \mathbf{P}_K + \frac{h}{2\pi i} \tag{9}$$

as an abbreviated form for the identity

$$\frac{h}{2\pi i}\frac{\partial}{\partial Q_K} Q_K \,.\, F(Q) = Q_K \frac{h}{2\pi i}\frac{\partial}{\partial Q_K} F(Q) + \frac{h}{2\pi i} F(Q).$$

The two equations (8) and (9) express only the statement that $\mathbf{p}_K$ shall stand for $\frac{h}{2\pi i}\frac{\partial}{\partial q_K}$ and $\mathbf{P}_K$ for $\frac{h}{2\pi i}\frac{\partial}{\partial Q_K}$. Another form of the same statement is that the probability amplitude $\psi_{p'}(q)$ is defined as a solution of the equations

$$(\mathbf{p}_K - p'_K)\,\psi_{p'}(q) = 0,$$

that is, of

$$\frac{h}{2\pi i}\frac{\partial}{\partial q_K}\psi_{p'}(q) - p'_K \,.\, \psi_{p'}(q) = 0,$$

and thus has the form

$$\psi_{p'}(q) = \text{const.}\, e^{\frac{2\pi i}{h}(p_1' q_1 + p_2' q_2 + \ldots)}; \tag{10}$$

and similarly that the probability amplitude $\psi_{P'}(Q)$ has the exponential periodic form

$$\psi_{P'}(Q) = \text{const.}\, e^{\frac{2\pi i}{h}(P_1' Q_1 + P_2' Q_2 + \ldots)}. \tag{11}$$

The exponential periodic form of the probability amplitudes $\psi_{P'}(Q)$ can be used conversely as a criterion for the "conjugacy" of the two physical quantities $Q_K(q,p)$ and $P_K(q,p)$ in quantum mechanics.

So we have altogether *three equivalent criteria* that the transformation q, $\mathbf{p}$ into $Q=Q(q,\mathbf{p})$ and $P=P(q,\mathbf{p})$ leads to new operators $\mathbf{P}$ "conjugate" to Q.

First, as a consequence of $\psi_{p'}(q)$ being described by the exponential function (10), it shall follow that the new amplitude $\psi_{P'}(Q)$ has the exponential form of (11).

Second, as a consequence of $\mathbf{p}$ and q satisfying exchange rules (8), it shall follow that the new $\mathbf{P}$ and Q satisfy the exchange rules (9).

Third, as a consequence of giving $\mathbf{p}_K$ the meaning of $\frac{h}{2\pi i}\frac{\partial}{\partial q_K}$, it shall follow that we must give $\mathbf{P}_K$ the meaning of $\frac{h}{2\pi i}\frac{\partial}{\partial Q_K}$.

Any of these three criteria may be used to test the "conjugacy" of the new system $Q(q,\mathbf{p})$, $\mathbf{P}(q,\mathbf{p})$.

§ 62. FIRST METHOD OF CANONICAL TRANSFORMATION

We introduced "contact transformations" q, $\mathbf{p}\to Q$, $\mathbf{P}$ by the definition (6). We wish to demonstrate that, by virtue of (6), at least *one* (hence all) of these three criteria for obtaining conjugate variables is satisfied. Take first as an example the case of a *point transformation*, where

$$Q_K=Q_K(q). \tag{12}$$

Here we had to take (§ 41) the operators $\mathbf{P}_K=\mathbf{P}_K(q,\mathbf{p})$ as solutions of the linear equations

$$\mathbf{p}_K=\sum_L \frac{\partial Q_L}{\partial q_K}\cdot\mathbf{P}_L. \tag{12'}$$

This transformation (12) and (12′) immediately satisfies the third criterion. Indeed, replacing in (12′) $\mathbf{P}_L$ by $\frac{h}{2\pi i}\frac{\partial}{\partial Q_L}$, we obtain, by operating on both sides upon the same function $f(q)=F(Q)$,

$$\mathbf{p}_K f(q)=\sum_L \frac{\partial Q_L}{\partial q_K}\frac{h}{2\pi i}\frac{\partial}{\partial Q_L}F(Q)=\frac{h}{2\pi i}\sum_L \frac{\partial F}{\partial Q_L}\frac{\partial Q_L}{\partial q_K}=\frac{h}{2\pi i}\frac{\partial f}{\partial q_K}.$$

Setting $\mathbf{P}_L=\frac{h}{2\pi i}\frac{\partial}{\partial Q_L}$, we must necessarily identify $\mathbf{p}_K$ with $\frac{h}{2\pi i}\frac{\partial}{\partial q_K}$, which is criterion 3. We may condense this demonstration to this sentence: One half of the system of equations (12) and (12′), namely the system (12′), is satisfied *identically* by virtue of the other half (12) if the symbol $\mathbf{P}_L$ is replaced by $\frac{h}{2\pi i}\frac{\partial}{\partial Q_L}$ and $\mathbf{p}_K$ by $\frac{h}{2\pi i}\frac{\partial}{\partial q_K}$. Hence (12) and (12′) give the transition from the conjugate system q, $\mathbf{p}$ into the new conjugate system $Q(q)$ and $\mathbf{P}(q, \mathbf{p})$.

What has been shown here for point transformations should now be demonstrated for general contact transformations as defined by (6). That is, it should be shown that one half of the system of the operator equations (6) is *identically* satisfied as a consequence of the other half of them, if $\mathbf{p}_K$ is given the meaning of $\frac{h}{2\pi i}\frac{\partial}{\partial q_K}$ and $\mathbf{P}_K$ the meaning of $\frac{h}{2\pi i}\frac{\partial}{\partial Q_K}$, *for any arbitrary form of the function* $S(Q, \mathbf{p})$. In order to facilitate this proof P. Jordan (21) supposes that the function $S(Q, \mathbf{p})$ has the form of a sum of products

$$S(Q,\mathbf{p})=\sum_n f_n(Q)\cdot g_n(\mathbf{p}), \tag{13}$$

so that the equations (6) have the form

$$\sum_n f_n(Q)\frac{\partial g_n(\mathbf{p})}{\partial \mathbf{p}_K}=q_K,\quad \sum_n \frac{\partial f_n(Q)}{\partial Q_K}\cdot g_n(\mathbf{p})=\mathbf{P}_K. \tag{13′}$$

But the proof is very complicated. It is all the more interesting that Jordan has found another method of transformation which evidently satisfies the *second* criterion, and thus leads from the conjugate system q, $\mathbf{p}$ to the new conjugate system Q, $\mathbf{P}$. This second method of transformation, which is, on the other hand, not so evidently in *correspondence* with the contact transformations of classical mechanics, will be described presently.

§ 63. SECOND METHOD OF CANONICAL TRANSFORMATION

Let $T(q,p)$ be any function of the q and p, and $T(q,\mathbf{p})$ its corresponding operator. The reciprocal operator $T^{-1}(q,\mathbf{p})$ is then defined as the operator that cancels the effect of T so that the successive application of T and T^{-1} is the multiplication by the factor unity

$$TT^{-1}=T^{-1}T=1. \tag{14}$$

If, for example, T has the form $T=\mathbf{p}_K=\frac{h}{2\pi i}\frac{\partial}{\partial q_K}$, then the reciprocal T^{-1} is the integral operator $\frac{2\pi i}{h}\int dq_K \ldots$.

Let $F(q,p)$ be any physical function and $F(q,\mathbf{p})$ its corresponding operator. We can then prove the formula

$$T(q,\mathbf{p})\,F(q,\mathbf{p})\,T^{-1}(q,\mathbf{p})=F(TqT^{-1},T\mathbf{p}T^{-1}). \tag{15}$$

The proof is based on the assumption that $F(q,\mathbf{p})$ can be expanded into a sum of products of the q's and $\mathbf{p}$'s. If for example $F(q,\mathbf{p})$ is only the one term $q_L\mathbf{p}_K^2 q_M$, then we have

$$\begin{aligned}TF(q,\mathbf{p})\,T^{-1}&=T(q_L\mathbf{p}_K^2q_M)\,T^{-1}=Tq_LT^{-1}T\mathbf{p}_KT^{-1}T\mathbf{p}_KT^{-1}Tq_MT^{-1}\\&=(Tq_LT^{-1})(T\mathbf{p}_KT^{-1})^2(Tq_MT^{-1})=F(TqT^{-1},T\mathbf{p}T^{-1}).\end{aligned}$$

If F is a sum of such products, the same proof applies to each of them and then to their sum as a whole. Call the right-hand side of (15), which is a function of the q and $\mathbf{p}$, $\mathscr{F}(q,\mathbf{p})$. Then we have

$$TF(q,\mathbf{p})\,T^{-1}=F(TqT^{-1},T\mathbf{p}T^{-1})=\mathscr{F}(q,\mathbf{p}). \tag{16}$$

Multiplying each term in front by T^{-1} and in back by T, we obtain the inverse formula

$$T^{-1}\mathscr{F}(q,\mathbf{p})\,T=\mathscr{F}(T^{-1}q\,T,T^{-1}\mathbf{p}T)=F(q,\mathbf{p}). \tag{16'}$$

Let us now introduce a number of given physical functions $Q_K(q,p)$ which we may call "new co-ordinates". Their corresponding operators are $Q_K(q,\mathbf{p})$. The method of finding their conjugate momentum operators $\mathbf{P}_K(q, \mathbf{p})$ runs as follows. First we try to find a transformer function $T(q, \mathbf{p})$ which has the property of satisfying the equations

$$(17) \qquad T^{-1} q_K T = Q_K, \quad \text{hence} \quad q_K = T Q_K T^{-1}.$$

(The actual process of finding such a transformation function is just as difficult as finding the operator $S(Q, \mathbf{p})$ applied in the first method of transformation.) Then we define the momentum operators by

$$(17') \qquad T^{-1} \mathbf{p}_K T = \mathbf{P}_K, \quad \text{hence} \quad \mathbf{p}_K = T \mathbf{P}_K T^{-1}.$$

This definition of "conjugate" operators $\mathbf{P}_K$ complies with the second criterion of § 61. Indeed we have

$$\mathbf{P}_K Q_K - Q_K \mathbf{P}_K = T^{-1} \mathbf{p}_K T \,.\, T^{-1} q_K T - T^{-1} q_K T \,.\, T^{-1} \mathbf{p}_K T$$
$$= T^{-1} (\mathbf{p}_K q_K - q_K \mathbf{p}_K) T$$

and if we suppose that $\mathbf{p}_K q_K - q_K \mathbf{p}_K = \frac{h}{2\pi i}$, we obtain as a consequence $\mathbf{P}_K Q_K - Q_K \mathbf{P}_K = \frac{h}{2\pi i}$, also. Thus (17) and (17′) indeed describe a transformation into conjugate Q_K, $\mathbf{P}_K$ if the original q_K, $\mathbf{p}_K$ are supposed to be conjugate.

If we insert (17) and (17′) in (16) and (16′) and read these equations from left to right, we obtain

$$(18) \qquad \mathscr{F}(q, \mathbf{p}) = F(TqT^{-1}, T\mathbf{p}T^{-1}) = TF(q, \mathbf{p}) T^{-1},$$

$$(18') \qquad F(q, \mathbf{p}) = \mathscr{F}(T^{-1}qT, T^{-1}\mathbf{p}T) = \mathscr{F}(Q, \mathbf{P}).$$

Thus after having found the transformer function T which mediates the transition to new co-ordinates Q_K according to (17) and to new momentum operators (18), we can find the expression $\mathscr{F}(Q, \mathbf{P})$ in terms of the new variables equal to a physical function $F(q, \mathbf{p})$ given in terms of the original variables, namely, we must form the operator $TF(q, \mathbf{p}) T^{-1} = \mathscr{F}(q, \mathbf{p})$ and replace the letters q and $\mathbf{p}$ by the letters Q and $\mathbf{P}$.

Of particular importance are the so-called *unitary transformations* whose transformation function $T(q, \mathbf{p})$ has the property that its *reciprocal* T^{-1} is identical with the complex conjugate of the *transposed* transformation function. That means, if $f(q)$ and $g(q)$ are any functions:

$$f T^{-1} g = g T^* f. \tag{19}$$

For other characteristics of unitary transformation functions see (32), § 66. The method of carrying out a transition $q, \mathbf{p} \to Q, \mathbf{P}$ by means of a "transformer function" T is preferable to the method using the "operator of action" $S(Q, \mathbf{p})$ as long as we wish to prove general theorems of quantum mechanics. But in all practical cases in which we really ask for the conjugates of given "new coordinates", the S-method is easier to carry out. We see this in the case of a point transformation, where S could immediately be written down (13), whereas T turns out to be a much more complicated expression. The general relation between the two operators $S(Q, \mathbf{p})$ and $T(q, \mathbf{p})$ which bring about the same transformation $q, \mathbf{p} \to Q, \mathbf{P}$ is given by the formula of P. Jordan:

$$T(q, \mathbf{p}) = e^{\frac{2\pi i}{h}\{S(q, \mathbf{p}) - \Sigma q_\kappa \mathbf{p}_\kappa\}},$$

where we have replaced $S(Q, \mathbf{p})$ in the exponent by $S(q, \mathbf{p})$. The operators now appear in the exponent, which means that we must expand the exponential function into a power series, $\mathbf{p}_K$ being treated like an ordinary quantity but obeying the exchange rules.

§ 64. PROOF OF THE TRANSFORMATION THEOREM

We wish to prove that the three solutions of (2), (3) and (4) satisfy the interference relation (1) supposing that the operators $\mathbf{P}_K(q, \mathbf{p})$ in $\mathrm{B}(Q, \mathbf{P}) = \beta(q, \mathbf{p})$ are conjugate to the $Q_K(q, \mathbf{p})$.

The transition $q, \mathbf{p} \to Q, \mathbf{P}$ may be described by the transformation formulae (17) and (17′). According to (18) $\mathrm{B}(q, \mathbf{p})$ is given by

$$\mathrm{B}(q, \mathbf{p}) = T \beta(q, \mathbf{p}) T^{-1}. \tag{20}$$

Then from (2) we have since $T^{-1} T = 1$

$$T^{-1} T \beta(q, \mathbf{p}) T^{-1} T \phi_{\beta'}(q) = \beta' . \phi_{\beta'}(q),$$

and because of (20)

$$T^{-1} \mathrm{B}(q, \mathbf{p}) T \phi_{\beta'}(q) = \beta' . \phi_{\beta'}(q).$$

Multiplying on the left by T, we obtain

$$\mathrm{B}(q, \mathbf{p})\, T\phi_{\beta'}(q) = \beta' \,.\, T\phi_{\beta'}(q).$$

Comparing this with (4), we obtain the equation

$$T\phi_{\beta'}(q) = \Phi_{\beta'}(q) \tag{21}$$

and its inverse $\quad \phi_{\beta'}(q) = T^{-1}\Phi_{\beta'}(q).$

These important equations show the relation between the two probability amplitudes $\Phi_{\beta'}(Q)$ and $\phi_{\beta'}(q)$ belonging to the same eigen-value β' but expressed in the two different co-ordinate systems q and Q.

If $G(Q)$ is any function of the Q's consisting of sums of products, it follows from (3) that not only

$$Q'\psi_{Q'}(q) = Q(q, \mathbf{p})\, \psi_{Q'}(q),$$

but also $\quad G(Q')\, \psi_{Q'}(q) = G(Q(q, \mathbf{p}))\, \psi_{Q'}(q),$

as can be seen by successive application of the examples $G(Q) = Q$, then $G(Q) = Q^2$, then $G(Q) = Q^3$, and so on. Owing to (16) we can write instead of the last equation

$$G(Q')\, \psi_{Q'}(q) = T^{-1} G(q)\, T\psi_{Q'}(q).$$

Integrating with respect to Q', we obtain

$$\int G(Q')\, \psi_{Q'}(q)\, dQ' = T^{-1} G(q) \int T\psi_{Q'}(q)\, dQ'. \tag{22}$$

The integrand $T\psi_{Q'}(q)$ vanishes, however, except for $Q' = q$, since according to (3)

$$T^{-1} q T\psi_{Q'}(q) = Q'\psi_{Q'}(q);$$

hence $\quad qT\psi_{Q'}(q) = Q'T\psi_{Q'}(q).$

Thus the last integral reduces to an integral over a range dQ' infinitely close to $Q' = q$ only, and has a *constant* value, independent of the value q. By means of a normalizing factor the constant value of the integral can be made *unity*. So (22) reads

$$\int G(Q')\, \psi_{Q'}(q)\, dQ' = T^{-1} G(q). \tag{23}$$

Now let us take for $G(q)$ in particular the function $\phi_{\beta'}(q)$, and we obtain from (21), (23)

$$\int \Phi_{\beta'}(Q)\,\psi_{Q'}(q)\,dQ' = T^{-1}\Phi_{\beta'}(q) = \phi_{\beta'}(q),$$

which is the equation of interference to be proved. The theorem then applies to the transition to any other set of conjugates $Q'(q, \mathbf{p})\,\mathbf{P}'(q, \mathbf{p})$ or $Q''(q, \mathbf{p})\,\mathbf{P}''(q, \mathbf{p})$. So the theorem is *invariant* with respect to the transition from one to another set of co-ordinates and conjugate operators $Q, \mathbf{P} \to Q', \mathbf{P}' \to Q'', \mathbf{P}''$ with the help of transformers $T', T'', \ldots$. This invariance corresponds to the invariance of the Hamilton equations of motion in classical mechanics to contact transformations (§ 49).

§ 65. INVARIANCE OF THE MATRIX ELEMENTS AGAINST UNITARY TRANSFORMATIONS

In § 48 we defined the matrix elements $[F]_{\beta''\beta'}$ of a physical function F with respect to the transition of another physical function $\beta(q, p)$ from the value β' to β'':

$$[F]_{\beta''\beta'} = \int \Psi^*_{\beta''}(F)\,\Psi_{\beta'}(F)\,F\,dF. \tag{24}$$

$[F]_{\beta''\beta'}$ gave the average of F in the state of transition $\beta' \to \beta''$, since $\Psi_{\beta'}(F)$ represents the probability amplitude of finding a particle with the value F in the pure case that its value $\beta(q, p)$ has the value β', and $\Psi^*_{\beta''}(F)\,.\,\Psi_{\beta'}(F) = \rho_{\beta''\beta'}(F)$ represents the transition density of a value F in the state of transition from the pure state β' to β''. We stated in § 48 without proof that the same matrix elements of the same function $F(q, p)$ can also be written in the form

$$[F]_{\beta''\beta'} = \int \phi^*_{\beta''}(q)\,F(q, \mathbf{p})\,\phi_{\beta'}(q)\,dq, \tag{25}$$

$F(q, \mathbf{p})$ operating on $\phi_{\beta'}(q)$, which is the probability amplitude of finding a particle at q when its property $\beta(q, p)$ has the value β'.

Finally, if other co-ordinates and conjugate momenta Q and P were introduced, and if $\Phi_{\beta'}(Q)$ was the probability amplitude for finding a value Q of the co-ordinate $Q(q,p)$ when $\beta(q,p)$ has the value β', we wrote

$$(26) \qquad [F]_{\beta''\beta'} = \int \Phi^*_{\beta''}(Q)\,\mathscr{F}(Q,\mathbf{P})\,\Phi_{\beta'}(Q)\,dQ,$$

supposing that $F(q,\mathbf{p})$ is transformed into $\mathscr{F}(Q,\mathbf{P})$ with the help of a *unitary* transformation (19), (17′). We now prove with F. London (22) that the three expressions for $[F]_{\beta''\beta'}$ are identical, in particular that (25) is identical with (26). The form (24) is then only a special case of (26), in which $F(q,\mathbf{p}) = \mathscr{F}(Q,\mathbf{P})$ is taken as one of the new co-ordinates Q itself and is called F. The identity of (25) and (26) expresses the *invariance of the matrix elements* against a unitary transformation from variables q, p to new conjugates Q, P. The identity of (25) and (26) is proved by the following succession of equations. Starting with (25) we have

$$[F]_{\beta''\beta'} = \int \phi^*_{\beta''}(q)\,F(q,\mathbf{p})\,\phi_{\beta'}(q)\,dq$$

$$= \int \phi^*_{\beta''}(q)\,T^{-1}TF(q,\mathbf{p})\,T^{-1}T\phi_{\beta'}(q)\,dq.$$

Using now $TFT^{-1} = \mathscr{F}$ according to (18) and $T\phi = \Phi$ according to (21), we obtain

$$[F]_{\beta''\beta'} = \int \phi^*_{\beta''}(q)\,T^{-1}\mathscr{F}(q,\mathbf{p})\,\Phi_{\beta'}(q)\,dq.$$

Supposing that T is *unitary* so that $fT^{-1}g = gT^*f$ according to (19), we obtain

$$[F]_{\beta''\beta'} = \int \mathscr{F}(q,\mathbf{p})\,\Phi_{\beta'}(q)\,T^*\phi^*_{\beta''}(q)\,dq,$$

and using (21) once more we have

$$[F]_{\beta''\beta'} = \int \mathscr{F}(q,\mathbf{p})\,\Phi_{\beta'}(q)\,\Phi^*_{\beta''}(q)\,dq.$$

Replacing the letters q, $\mathbf{p}$ in the integrand by the letters Q, $\mathbf{P}$, we obtain (26), and the invariance is thus proved.

§66. MATRIX MECHANICS

For the sake of completeness we shall derive briefly some interesting properties of the matrix mechanics. First we see immediately that the matrix elements are the expansion coefficients in a series with respect to the eigen-functions

$$(27) \qquad F(q, \mathbf{p})\,\phi_{\beta'}(q) = \sum_{\beta''} [F]_{\beta''\beta'}\,\phi_{\beta''}(q).$$

This equation is proved by multiplying (27) by the complex conjugate of one of the eigen-functions, for instance by $\phi^*_{\beta'''}$, and integrating with respect to dq and using the orthogonality and normalization of the eigen-functions

$$\int \phi^*_{\beta'''}\phi_{\beta''}\,dq = 1 \text{ for } \beta''' = \beta''$$
$$= 0 \text{ for } \beta''' \neq \beta''.$$

Furthermore, we see immediately that the matrix element of the sum of two functions $G(q, p)$ and $F(q, p)$ is the sum of their matrix elements

$$(27') \qquad [F+G]_{\beta''\beta'} = [F]_{\beta''\beta'} + [G]_{\beta''\beta'} \quad \text{(rule of addition)}.$$

The product of two functions $F(q, p)$, $G(Q, p)$, however, has matrix elements that are composed of the matrix elements of F and those of G in a more complicated fashion. On the one hand, according to (26), we have

$$(28) \qquad F(q, \mathbf{p})\,G(q, \mathbf{p})\,\phi_{\beta'}(q) = \sum_{\beta''} [FG]_{\beta''\beta'}\,\phi_{\beta''}(q).$$

On the other hand, we can write

$$(28') \quad F(q, \mathbf{p})\,G(q, \mathbf{p})\,\phi_{\beta'}(q) = F(q, \mathbf{p}) \sum_{\beta'''} [G]_{\beta'''\beta'}\,\phi_{\beta'''}(q)$$
$$= \sum_{\beta'''} [G]_{\beta'''\beta'}\,F(q, \mathbf{p})\,\phi_{\beta'''}(q) = \sum_{\beta'''} [G]_{\beta'''\beta'} \sum_{\beta''} [F]_{\beta''\beta'''}\,\phi_{\beta''}(q)$$
$$= \sum_{\beta''} \Big(\sum_{\beta'''} [F]_{\beta''\beta'''}\,[G]_{\beta'''\beta'}\Big)\,\phi_{\beta''}(q).$$

Equating the right-hand sides of (28) and (28′):

$$\sum_{\beta''} [FG]_{\beta''\beta'}\,\phi_{\beta''}(q) = \sum_{\beta''} \Big(\sum_{\beta'''} [F]_{\beta''\beta'''}\,[G]_{\beta'''\beta'}\Big)\,\phi_{\beta''}(q),$$

the factors of each single $\phi_{\beta''}(q)$ in the sums must be equal:

$$(29)\qquad [FG]_{\beta''\beta'} = \sum_{\beta'''} [F]_{\beta''\beta'''} [G]_{\beta'''\beta'} \quad \text{(rule of multiplication)}.$$

(29) shows how the matrix elements of a product function FG are composed from those of the single functions F and G. (29) has the same form as the rule according to which the elements $F_{\beta'\beta''}$ and $G_{\beta'\beta''}$ of two determinants $|F|$ and $|G|$ are composed to form the elements of the product determinant $|FG|$, multiplying lines by columns. We notice that the matrix elements of FG are different from those of GF:

$$(29')\qquad [GF]_{\beta''\beta'} = \sum_{\beta'''} [G]_{\beta''\beta'''} [F]_{\beta'''\beta'}.$$

In particular, if $F(q, \mathbf{p}) = q$ and $G(q, \mathbf{p}) = \mathbf{p}$ we can ask for the difference of the matrix elements of pq from those of qp; we obtain here

$$(30)\qquad [\mathbf{p}q]_{\beta''\beta'} - [q\mathbf{p}]_{\beta''\beta'} = \int \phi^*_{\beta''} (\mathbf{p}q - q\mathbf{p})\, \phi_{\beta'}\, dq.$$

Since the operator in the integrand is equivalent to a multiplicand $h/2\pi i$ according to (8), we obtain on the right

$$\int \phi^*_{\beta''} \frac{h}{2\pi i} \phi_{\beta'}\, dq = \frac{h}{2\pi i} \text{ for } \beta'' = \beta',$$
$$= 0 \quad \text{for } \beta'' \neq \beta'.$$

Although the products pq and qp are identical, their matrix elements are different, their difference being

$$(31)\qquad [\mathbf{p}q]_{\beta''\beta'} - [q\mathbf{p}]_{\beta''\beta'} = \frac{h}{2\pi i} \text{ for } \beta'' = \beta',$$
$$= 0 \quad \text{for } \beta'' \neq \beta'.$$

(31) represents the Heisenberg exchange rule for the matrix elements of qp and pq which is also responsible for the difference of (29) from (29′). Of special interest are the matrix elements of a "unitary" function $T(q, p)$ as defined in (19). If we introduce the symbol $[\widetilde{T}]_{\beta''\beta'}$ for $[T]_{\beta'\beta''}$ with transposed indices, then we have

$$[\widetilde{T}]_{\beta''\beta'} = [T]_{\beta'\beta''} = \int \phi^*_{\beta'}(q)\, T(q, \mathbf{p})\, \phi_{\beta''}(q)\, dq$$

and its complex conjugate

$$[\tilde{T}]^*_{\beta''\beta'} = \int \phi_{\beta'}(q)\, T^*(q, \mathbf{p})\, \phi^*_{\beta''}(q)\, dq.$$

Using (19), we can continue,

$$= \int \phi^*_{\beta''}(q)\, T^{-1}(q, \mathbf{p})\, \phi_{\beta'}(q)\, dq = [T^{-1}]_{\beta''\beta'}$$

with the result

(32) $$[T^{-1}]_{\beta''\beta'} = [\tilde{T}]^*_{\beta''\beta'} = [T]^*_{\beta'\beta''}.$$

In words: If we take the complex conjugates of the transposed matrix elements of a unitary operator $T(q, \mathbf{p})$ we obtain the matrix elements of its reciprocal operator $T^{-1}(q, \mathbf{p})$.

Since the operator $T(q, \mathbf{p})\, T^{-1}(q, \mathbf{p})$ is equivalent to a multiplication by unity, we obtain the matrix elements

$$\begin{aligned} [TT^{-1}]_{\beta'\beta''} &= 1 \text{ for } \beta' = \beta'', \\ &= 0 \text{ for } \beta' \neq \beta''. \end{aligned}$$

If T is unitary, we obtain then, at the same time,

(33) $$\begin{aligned} [T\tilde{T}^*]_{\beta'\beta''} &= 1 \text{ for } \beta' = \beta'', \\ &= 0 \text{ for } \beta' \neq \beta'', \end{aligned}$$

which means, according to (29),

(33′) $$\begin{aligned} \sum_{\beta'''} [T]_{\beta'\beta'''}\, [T]^*_{\beta''\beta'''} &= 1 \text{ for } \beta' = \beta'', \\ &= 0 \text{ for } \beta' \neq \beta''. \end{aligned}$$

These equations are quite analogous to the relations between the directional cosines in an orthogonal transformation which transforms a *unit* vector into another unit vector of different direction:

$$\sum_k \alpha_{mk}\alpha_{nk} = \begin{matrix} 1 \text{ for } m = n, \\ 0 \text{ for } m \neq n. \end{matrix}$$

This analogy accounts for the term "unitary" used for transformation (19). Owing to the operator transformations (17), (17′), we have the matrix rules

$$[Q_K]_{\beta''\beta'} = [T^{-1} q_K T]_{\beta''\beta'},$$

or, owing to the product rule (29),

$$[Q_K]_{\beta''\beta'} = \sum_{\beta'''} \sum_{\beta^0} [T]^*_{\beta''\beta'''}\, [q_K]_{\beta'''\beta^0}\, [T]_{\beta^0\beta'},$$

and because of (32),

$$(34)\qquad [Q_K]_{\beta''\beta'} = \sum_{\beta'''}\sum_{\beta^0} [T^{-1}]_{\beta''\beta'''} [q_K]_{\beta'''\beta^0} [T]_{\beta^0\beta'},$$

and the corresponding formula (34′) for the matrix elements of the conjugate momenta P_K. The great advantage of unitary transformations is that we have nothing to do with the reciprocal operator T^{-1} or its matrix elements, T^{-1} being replaced by $\widetilde{T}^*$ and $[T^{-1}]_{\beta''\beta'}$ by $[T]^*_{\beta'\beta''}$.

Upon the rules of addition (27′) and multiplication (29) and on the exchange rule (31) it is possible to build up the calculus of *matrix algebra*. The transformation formulae (34), (34′) and (33′) then lead to the branch of quantum mechanics which has been developed by Born, Heisenberg and Jordan. Its chief concern is the direct calculation, without resorting to unobservable complex probability amplitudes, of the observable averages of physical functions $Q(q,p)$ represented by their matrix elements in various states of transition from one value β'_K to another value β''_K of other physical quantities $\beta_K(q,p)$.

In all physical applications the $\beta_K(q,p)$ are assumed to be constants of the motion possessing eigen values β'_K, β''_K, ... The matrix elements $[Q]_{\beta'\beta''}$ represent the values of a physical quantity Q if the latter is interpreted by means of the wave theory of observation, in a state which is described in corpuscular terms as a sudden transition between the constant values β'_K and β''_K of the quantities $\beta_K(q,p)$. Only if the $\beta_K(q,p)$ are constants of the motion does there exist a correspondence between classical and quantum mechanics (§ 52 and § 58), although the mathematical calculus of quantum mechanics can be carried through regardless of this physical restriction.

INDEX OF LITERATURE

(1) L. de Broglie, Thèse, Paris, 1924; *Ann. de Phys.* sér. 10, **3**, p. 22, 1925.

(2) E. Schrödinger, *Ann. Phys.* **79**, pp. 361 and 489, 1926.

(3) M. Born, *Zs. f. Physik* **38**, p. 803, 1926; P. A. M. Dirac, *Proc. Roy. Soc.* **113**, p. 621, 1926.

(4) W. Heisenberg, *Zs. f. Physik* **33**, p. 879, 1925.

(5) M. Born and P. Jordan, *Zs. f. Physik* **34**, p. 858, 1925; **35**, p. 557, 1926.

(6) P. A. M. Dirac, *Proc. Roy. Soc.* **109**, p. 642, 1925; **110**, p. 561, 1926.

(7) N. Bohr, *Naturwiss.* **16**, p. 245, 1928; **17**, p. 483, 1929; **18**, p. 73, 1930.

(8) W. Duane, *Proc. Nat. Acad. America* **9**, p. 158, 1923.

(9) P. A. M. Dirac, *Proc. Roy. Soc.* **114**, p. 243, 1927.

(10) W. Heisenberg, *Zs. f. Physik* **43**, pp. 172 and 809, 1927.

(11) W. Heisenberg, *The physical principles of quantum theory,* University of Chicago Series, 1930.

(12) P. S. Epstein and P. Ehrenfest, *Proc. Nat. Acad. America* **10**, p. 133, 1924; **13**, p. 400, 1927.

(13) O. Halpern, *Zs. f. Physik* **30**, p. 153, 1924.

(14) W. Bothe and H. Geiger, *Zs. f. Physik* **32**, p. 639, 1925.

(15) A. H. Compton and A. W. Simon, *Phys. Rev.* **26**, p. 289, 1925.

(16) M. von Laue, *Ann. d. Physik* **44**, p. 1197, 1914.

(17) J. H. Jeans, *Phil. Mag.* **10**, p. 91, 1905.

(18) D. R. Hartree, *Proc. Cambr. Phil. Soc.* **24**, p. 89, 1928.

(19) E. Madelung, *Zs. f. Physik* **40**, p. 322, 1926.

(20) P. Ehrenfest, *Zs. f. Physik* **45**, p. 455, 1927.

(21) P. Jordan, *Zs. f. Physik* **38**, p. 513, 1926; *Göttinger Nachr.* p. 161, 1926.

(22) F. London, *Zs. f. Physik* **40**, p. 193, 1926.

INDEX OF NAMES AND SUBJECTS

Printed in the United States
By Bookmasters